The Mindlike Machine

A Universal Theory of Everything

T G Henderson

Copyright © 2024 T G Henderson
Lincolnshire, UK

https://timhenderson.uk ISBN: 9781068717918

All rights reserved. No part of this publication may be reproduced, distributed, or transmitted in any form or by any means, or stored in a database or retrieval system, without the prior written permission of the publisher.

Edition: 1.0
10th July 2024

For those who value truth and reason as the highest principles.

With gratitude to the people who made it possible.
My parents Phil and Gina, my brother James, my son Zachary.

Chapter 1

Introduction

A Universal Theory of Everything

Despite considerable efforts by many great minds over the years there is no sign that a substantial "Theory of Everything" (TOE) is about to emerge from physics. Whether one will ever be found, or is even possible, remains an open question.

Some believe it is an unachievable goal due to the implications of Gödel's incompleteness theorem. Even Steven Hawking, whose life was dramatized in the 2014 film "The Theory of Everything", eventually concluded that one was not possible for this reason.

"Some people will be very disappointed if there is not an ultimate theory that can be formulated as a finite number of principles. I used to belong to that camp, but I have changed my mind."

`http://yclept.ucdavis.edu/course/215c.S17/TEX/GodelAndEndOfPhysics.pdf`

This book claims that a TOE is possible and describes it in detail. However, it is not primarily a theory of physics, as Prof. Hawking might have imagined, but a truly universal one capable of describing all types of phenomena including "mind", "logic", and "information".

Far from being impossible, or even particularly hard to find, the basic principles are somewhat self-evident, "hidden in plain sight" within the structure of information itself. It seems to have been overlooked, not because it's too complicated for the human mind to comprehend, but rather because it's too simple and its significance was missed.

Our route to understanding the physical world is not direct, but indirect, via our senses and thoughts. To understand the outside, we must first understand the inside, the world of concepts. We "understand" things via "descriptions" which are "ideas"; all knowledge is made of ideas. (Note: I'll be using "quotes" to mean "in the most general sense".)

If we examine "ideas" we find there is an underlying natural structure from which they emerge. This structure is the foundation of information, thought, logic and mathematics, and hence all conceivable things. We find a system which is generalised but strict, capable of defining all things precisely.

It will be argued that the theory described in this book is the only conceivable TOE, the only possible logically consistent explanation of reality. It may not be the answer anyone wanted, but it is probably the right one.

Non-Physical Reality

In contrast to the assumption of materialism which underlies much of modern science, this theory is founded on the assumption of idealism, that this is a simulated universe made of information, not matter.

Idealism proposes that matter is not fundamental, and I will argue it is generally equivalent to simulation theory. Either the universe is made of matter or information. These are the only two categories of "stuff" that have structure and can be built with.

Simulation theory is widely recognised as being a plausible explanation for our existence. As computer simulations of virtual worlds become increasingly sophisticated, the proposal that we exist within a simulation seems ever more reasonable.

Most religions view the universe as a deliberate construction and as an "illusion" or "sub-reality" of some sort. Any concept of a "Heaven" or "higher plane" of existence that exists "above" this one can, in the most general sense, be considered as being "outside the simulation".

A theory of information

This theory can be described in many ways, but the simplest might be as a "theory of information".

If this universe is "made of data", then the "laws of the universe" are the same thing as the "law of information", and that is simply "logic". Logic defines which ideas are possible and can exist, and which are paradoxes and cannot. It defines the structure of information, which is "everything".

The proposal that information has an underlying natural structure shouldn't be particularly controversial. Logic obviously contains structure. It is also generally equivalent to "language", which is highly structured. A universe made of information would necessarily be constructed from some form of language, using logic.

Even if the assumption of idealism is incorrect, and universe is made of matter, we can only understand matter via information, so we need to know how it works anyway. We need to understand how we understand things, and this theory at least helps us accomplish that.

Whether the structure of reality matches the structure of information exactly is something that can only be determined by experiment, but if this is a simulation the two should be identical.

In summary, if this universe is fundamentally non-physical then we should expect that a "theory of information" would also be a theory of everything. This book proposes that this is indeed the case. It is possible to formulate a TOE via this route and it's relatively easy to do. It appears to work well, providing improved definitions for important concepts, diverse new theories, and strong testable predictions.

The Universal Plan

The system described I call the "Universal Plan" (UP), but it has clearly been discovered before, prior to current recorded history. There are "echoes" of it still existing in the modern day in various places, but the underlying structure itself was apparently lost, until now. It wasn't the intent, but this investigation also appears to have been a rediscovery from first principles of what was probably the original version of "Alchemy".

The UP is not just a philosophical theory, it has immediate practical utility. It can be applied as a method of analysis to discover how any natural system works at the most fundamental level. The book includes several examples of how this works in practise.

The theory proposes that the universe is made of "language", but also that the idea of "language" is generally equivalent to "relationships", "information" and "mind". It describes a new universal model for language with a new system of word classes which allows a unification between human and computer languages, demonstrating that they share the same basic form. The proposal is that every phenomenon in nature shares this same basic structure, and we can correlate them all via this method.

At the end of the book, I apply the theory to the "semi-physical" phenomenon of colour. The results are compelling with the system boldly predicting a new four-primary model of colour vision. It suggests the eye works via a completely different principle than is currently assumed, and its predictions match observation better than trichromatic theory, potentially explaining many unanswered questions.

Because the UP is a generalised qualitative theory, this investigation touches on many topics. It makes significant predictions, each of which deserves more detail than I'm able to go into here. These sub-theories really need a more rigorous explanation, suitable for scientific journals. Hopefully that will happen in due course.

New theories	Description
A new theory of information and language	A universal structure for all types of language.
A computational theory of mind	The "Qualia Processor".
A unified aether model of physics	A unification of light, energy and matter. A new fundamental model of physical matter. An asymmetrical model of charge. Magnetism explained.
A new theory of entropy and order	A single principle which can account for all physical structure.
New theories of colour and colour vision	A four primary colour model of the eye.
A new model of the prism	Interference is the only mechanism of colour production
A new theory of the senses	There are only four fundamental physical senses.
A new theory of categories (philosophy)	A definitive ontology of fundamental categories.
A new theory of morality	A simple, natural explanation of morality.
A new theory of free will	A full definition of the principles / archetypes of choice.
A new perspective on time	"Time" comes in two forms.
A new understanding of space / time	Time is the dual opposite of space.

Improved definition for crucial concepts	"logic", "information", "infinity"
The "Philosopher's Stone"	The rediscovery of Alchemy from first principles.
Proof of "God"	Logical proof of an intentional universal creator.

The Universal Plan (UP) is the foundation of a TOE compatible with the view of philosophical idealism, as it proposes the universe is fundamentally made of "ideas", not "matter". (Although, strictly speaking, ideas are a form of "matter".)

It's a "universal" theory in three senses.

- It applies universally to all things.
- It's about the universe. It explains what it is and how it works.
- It views reality as made of "universals", abstract-concepts, information.

universal
- applicable or occurring throughout or relating to the universe.
- term applied to general or abstract objects such as concepts, qualities, relations, and numbers, as opposed to particular objects.

The purpose of this book is to give a concise overview of the basic elements of the theory, to provide a starting point for discussion. It's intended to make the argument as simple, clear, and unambiguous as possible, and to provide a foundation for further work. It can't answer all possible questions, objections, or implications, it'd be too long. I've kept it as short as possible and left out a lot of discussion because that is best dealt with after the whole theory has been explained.

The concepts that emerge from the investigation offer an unusual perspective on the world. It leads to an entirely different paradigm of thought, a new way to view reality. Because of this, it may not be immediately obvious what information I'm attempting to convey, and the mode of thinking may be unfamiliar. Hopefully all will be clear by the end though.

A theory of everything will inevitably be extremely generalised. It has to reveal commonality between the most disparate things imaginable, that is its purpose. It must somehow bring everything that exists together into one by finding what all things in the universe have in common.

This theory requires conceptually "zooming-out" of reality to its most fundamental, lowest-detail level. It involves thinking about things in the most generalised way possible, about archetypes and symbols, and very broad concepts. It does so in order to remove as much vagueness and ambiguity as possible from them, but it may take a while for the picture to emerge.

We need to find ways to talk about extremely generalised ideas with a reasonable amount of precision, and it may not always be easy to find the right balance. There's a well-defined structure underlying the theory but finding the right words to explain the fundamental concepts can be tricky.

This is only the first edition, and while I believe the basics of the theory are sound, it's still a work in progress and there's plenty left to figure out. Feedback is welcome.

Note, there is no mathematics in this book. The universe is made primarily from qualities and numbers are always a "measurement" of those, they are a secondary phenomenon. Mathematics is a derivative of the UP and all quantities are derived from qualities, as will be explained.

Assumptions

To undertake this investigation, it's necessary to make two basic assumptions.

1. Reality is knowable.

We must assume the universe operates according to a logical set of rules which we can discover. This assumption is the foundation of traditional science, so shouldn't pose too much of a problem.

- All things have a cause. All causes have an effect.
- The universe follows logical, consistent, knowable laws.
- We can discover those laws by observation and experiment.

If any part of reality operates outside cause-and-effect, or works by any non-logical mechanism, that would put it beyond our understanding. We must assume everything occurs by logical processes because logic is the only tool we have for understanding things.

Some interpretations of quantum mechanics say things can happen without a cause, but this theory rules it out as being conceptually impossible. Some suggest that causality is not fundamental, but this also leads to paradoxes, so again is ruled out by the UP.

2. Reality is made of "ideas" / "information".

We will assume that the reality we experience isn't made of physical matter but of "information". This assumption is equivalent to assuming it is a "simulation" of some sort.

From First Principles

Other than these initial assumptions the investigation relies only on simple evidence and concepts which are as close to being "objective" as possible. The original idea was to explore only basic concepts and logic, and to see what emerged naturally. It turned out that was sufficient.

To understand this theory, it's helpful to try to start out with a blank slate. Some "out of the box" thinking is needed, but the theory relies only on information it's possible to personally know, first-hand.

To do any kind of analysis we need two fundamental abilities.
- The ability to observe.
- The ability to reason.

Observation and reason are the first principles of all knowledge and science. We can only truly know things by observing and thinking about them ourselves, and everything discussed here can be "observed" first hand (i.e. with the "mind's eye").

The theory is founded on observations anyone can make, and logic anyone can follow. The reader can personally verify every part of the argument, although if you want to verify the part on colour, you'll need a prism too.

The section on language and word-classes is a bit complicated, but it is still simpler than the current paradigm because it has a clearly defined structure.

The "Mind's Eye"

We can view reality as having two basic parts. There's an "outside world" of solid, tangible things made of matter, and an "inner world" made of intangible non-material things like ideas and emotions.

The "mind's eye" is an "internal sense", or set of senses, we use to observe non-material things like ideas. When we imagine things, make plans, or try to solve problems, we are using the mind's eye for that work.

One of the main aims of this book is to reframe certain key concepts, which are currently considered "mystical" and "unscientific", into a down-to-earth form which people are familiar with. This means they can be objectively described, bringing them into the realm of science.

The "third eye" is the "mind's eye".

I suggest what we call the "mind's eye" in the west is identical to the concept of the "third eye" in various religious traditions. Thus, the idea is not at all "mystical", but somewhat ordinary. The "opening of the third eye" is then equivalent to contemplation, thought, thinking for yourself. It is the stage of life where we start to think for ourselves and become "intellectually independent".

There is no need for some extra facility that a magical third eye might offer because the mind's eye lacks for nothing. It is perfectly capable of answering all questions on its own, no magic is required.

While physicists investigate the structure of matter to try to understand reality, idealists must study the structure of ideas and mind to achieve the same goal. Physicists look outward into the world of matter with physical eyes, idealists look inwards to the world of thought with the mind's eye.

"All Models Are Wrong"

Some believe a TOE is impossible.

"A number of scholars claim that Gödel's incompleteness theorem suggests that any attempt to construct a theory of everything is bound to fail."

`https://en.wikipedia.org/wiki/Theory_of_everything`

I'm not sure how Gödel's theorems might be relevant to the UP, although it may be that the UP's improved(?) definition of "infinity" obviates any problem. It's a complicated issue that's outside the scope of this book.

The concept of "paradox" is something the UP handles gracefully. There is a defined (and prominent) place for paradox in the framework, which means the system can contain it while remaining logically consistent.

This theory suggests that the paradoxes presented by Gödel may be category errors, as this seems to be what all such conflicts are, although more work is obviously needed on this. Certainly, some of the examples given of it are paradoxical (i.e. correlating real numbers with integers, which is impossible).

Some would contend that there is no right answer to the question we're asking, no perfect model.

"A common aphorism in statistics is "All models are wrong, but some are useful". This acknowledges that statistical models always fall short of the complexities of reality but can still be useful nonetheless. The aphorism originally referred just to statistical models, but it is now sometimes used for scientific models in general."

https://en.wikipedia.org/wiki/All_models_are_wrong

While this may be true in statistics, it's not always the case. Many software applications model real world processes perfectly. An online shopping cart models a physical one and is arguably much easier to use. Logical / mathematical models can fully describe reality because the foundation of reality is logic.

An Apology

I apologise in advance for any errors or other failings in this book, it is self-published. So, I ask the reader for a degree of flexibility and patience.

I am not an academic. I have a degree in biology, but my career has been in software. I'm sure there are, especially from an academic perspective, many imperfections in this book. It is intended for a more general audience, and it also needed to be simple enough to finish.

I would ask the reader to forgive any lack of academic prowess as a theory like this is unlikely to have ever come from academia. It's a long way from the current paradigm which doesn't naturally lead this way.

The UP offers a completely different perspective on reality, but that is inevitable given the topic. It is necessarily a new perspective on the universe and the framework of thought itself. It is a new paradigm, a "new mind". It is simply not possible to get to it from within the current academic framework, because it is an alternative framework.

I'd suggest that the only place this theory could have come from is outside academia.

As a non-academic, my career doesn't depend on how well these ideas are received by the world. I am free to propose them without fear of losing professional credibility, or my salary. It doesn't really matter if this book is universally condemned, ridiculed, or ignored; but I do hope it reaches those who might benefit from it, however few or many that may be.

I'm sure this book could have been written in many other ways, most of which would probably have been better, but this is the version we're stuck with for now. I hope it's good enough.

Overview

This theory claims that there is a universal, foundational, conceptual framework which exists in nature independent of the minds of mankind (the UP). It's a hierarchical system of broad archetypes, beginning with the most generalised concept(s) and progressing to more detail over successive levels.

This framework derives directly from the fundamental principles of logic and is a necessary construct that could not have been any other way. It is the natural structure of logic, information, language, thought and everything else. It's a "map of logic".

Reality is constructed from this "toolkit" of foundational ideas, it's like the "operating system" of the universe. The rules of mathematics and physics originate within this framework. Its ten different parts are the most fundamental components of the reality we perceive, and this arguably must be so for any conceivable world.

Ideas contain their own intrinsic logic and facts. If we analyse the contents of the most basic concepts, we find they have a story to tell. The world of ideas has its own strict order, and we can observe natural relationships within it. Ideas explain each other. The conceptual framework is self-referential and self-documenting.

The UP only deals with concepts. It defines what is conceptually possible, *i.e.* what ideas are logically consistent. The question of how well it matches physical reality remains to be determined. The evidence so far suggests it matches well but it's still early days for the theory. It makes many bold predictions which are testable, so there are plenty of routes available for validation / falsification.

Extreme Abstraction

I would ask the reader to bear in mind that the fundamental concepts described in this book are the most abstract ones that exist. They are the broadest ideas it is possible to think and are more like extremely generalised categories. They can be difficult to understand for this reason.

However, they are very simple at heart. If the reader can overcome the initial hurdle of the deeply abstract nature of the archetypes being discussed, then everything should eventually make sense.

A Universal Language

The UP framework takes the form of language. It is a "super-language", the "prototype" or "template" of information, and it can be read as a story of how to create something. It seems to be the simplest conceivable way to create a universe from ideas, and nature always prefers the simplest solutions.

The purpose of language is communication, to allow concepts to be shared with other minds. If the universe is structured like language, then it is a form of communication, and we may be the "other minds" it is trying to communicate with.

The UP seems to be designed to be understood. It's so (relatively) simple to understand, so accessible to inquiry, it's hard not to reach the conclusion that the universe is designed to be understood by us, to explain something. What is it trying to explain?

The "God" concept

The system itself provides an answer. It begins with the concept of "one", which we find correlates with "consciousness". Consciousness is "the individual", it is "to be". "One exists" is the same as "I am". "One" is a super-concept. It contains all other concepts within it; it is "everything", "the whole", and therefore "God".

The universe exists to explain "One", to give it context, *i.e.* something to compare itself against.

The universe can be viewed as an ongoing explanation of (at least) the following concepts.
- What is "One"? The character of the first concept, "One" / "God".
- Being and consciousness. What it is to exist.
- Unity and division. Being "one", or "many".

The concept of "God" carries an enormous amount of baggage in the modern world and is possibly the most emotive subject of all. As difficult as it might be, to truly understand the UP, it's necessary to leave all of that behind and approach the topic with an open mind.

The benefits of the UP to the religious community are many, as the UP explains the nature of "God" with more depth and precision than any other system and it proves the concept's necessity. However, that exhaustive description does not concur exactly with any individual modern belief system.

The UP says God must exist (at least conceptually), but also that God is not as is commonly described.

Observer / Observed

At the top level, we can view the framework as consisting of two basic ideas, observer and observed.

1. "One": The observer / creator

Consciousness. "I am".

One is the subject of discussion. The thing that needs to be described. It is the "individual", the "soul" (sole, only). It is the top-level of the hierarchy, the origin of all concepts, the "creator of everything".

2. "Many": The observed / creation

Matter originates in duality / information. This is "creation", the "tree of life", the universe.

The following table shows the three most fundamental archetypes, and top two ("Heaven") levels of the UP. The system is somewhat like a "fractal" with lower levels being a finer description of those above. As all levels refer to the top category, there is self-similarity between them.

Level 1 "One"	One, unity, single, alone, consciousness The original observer, "outside" the universe. Observing all that is below (inside). The container *host* creator. "To be" / "I am"	
Level 2 "Many"	The observed. The universe, "Yin" Matter, solid objects, data, information, nouns "I …"	The observer in the universe: "Yang" Spirit, abstract objects, instructions, verbs "… am"

There are two observer / observed relationships embedded in these three concepts, two different perspectives on reality. One observes creation as a whole "from the outside", Yang observes it from the inside.

We can demystify this relationship a bit more by comparing it to the following language construct:

"The dreamer (One) dreams (Yang) the dream (Yin)".

This archetypal pattern of three words is the underlying form for the most basic sentence. It can be rephrased more generally as "I do stuff". The words are representations of these three most fundamental archetypes.

I (One, consciousness) do (Yang, action) stuff (Yin, matter).

One: Consciousness, The Source

All concepts originate in the quality of "one" / unity. This is the root of the tree. It is a thing, an "object", a "noun". It's a non-physical entity that can be described. The idea of "one" is correlated with the following concepts.

One: single, singularity, unitary, unified, (the) whole, everything, consciousness, sole, only, alone.

The system starts with the most generalised concept possible, "everything", then that gets "divided down" into more and more detailed concepts as it descends through the levels. The framework begins as one thing with a single perspective, which then becomes many things with many different perspectives, via the process described.

While materialism says everything is "built up" from matter, the UP says matter is "divided down" from everything. These views are a natural duality.

"One" is the root of the language-tree and is the subject of it. The tree exists to explain it because it has no descriptive power on its own. This is a crucial observation:

You can't explain anything with just the concept of "one".

To describe things, we need to be able to compare them to something else. The concept of comparison is of great importance, closely linked to the idea of "description".

All descriptions are comparisons.

Duality: Observer and Observed

Level two of the system is the number "two", duality. Again, taken as a whole, this is an object / noun like entity.

Two: duality, many, separate, divided, difference, comparison, conflict, cooperation.

One is unified, two is divided. Duality is the opposite of unity, and it is the origin of all information and "meaning". It enables us to begin to describe things, albeit in a very generalised way. This is the one-dimensional level of the hierarchy. ("One" has no dimensions.)

Duality is a complex idea. It inherently contains the concept of logic, and it hosts two distinct opposed sub-categories which are the foundation for all types of description. They are "Yin and Yang" (I'll explain why).

Yin and Yang are the most fundamental "properties" that exist.

These two sub-categories contain (or "divide into") lists of properties which describe One. Yang contains a list of properties that One has, Yin contains a list of properties One does not have.

All conceivable fundamental properties are dualities, and they all contain two parts which can be categorised into "unitary or multiple", "one or many", Yang or Yin. *E.g.*

One: first, leader, single, one, direct, straight, near... (Yang +)
Two: second, follower, multiple, many, indirect, curved, far... (Yin -)

There is natural correspondence between the properties in a category, they are all related. If something is Yin in a relationship, it will have all the context-suitable Yin properties. They come as a package. This provides powerful predictive abilities to the theory.

Duality gives rise to seven more archetypes, and that completes the set of ten.

Ten "Super-Concepts"

There are ten distinct entities in the system. The following table shows what the UP looks like as a whole, as applied to human language (HL) and computer language (CL). It shows the relationships between all the concepts, and how they descend through four levels. It demonstrates that all languages are analogous, and have the same fundamental parts, corresponding to this structure.

Because of the self-referential, "fractal" nature of the system, the top two levels are categorised as Yang ("Heaven" or "Sky"), and the lower two as Yin ("Earth"). Yang is simple *spirit, Yin is complex* matter.

Level 1. One, unity HL: "To be" / "I am" CL: The executive: "run", "exec"...			
Yin − Matter, solid objects HL: Noun CL: Data		**Yang + Spirit, abstract objects** HL: Verb CL: Instruction	
Earth -- HL: Determiners CL: Assignments	**Water -+** HL: Adjectives CL: Functions	**Air +-** HL: Adverbs CL: Comparison	**Fire ++** HL: Questions CL: Flow control
Sex HL: Prepositions CL: Math operators	**Heart** HL: Time-Joiners CL: Code-blocks	**Voice** HL: Conjugations CL: Logic operators	

Seven Principles

The table below shows the final level of the system, the set of seven principles in the lower two levels above.

These are the "Earth" archetypes. They have the same (general) form as the human body. They also neatly correspond to the question words ("interrogatives").

Archetype	Includes concepts like...	Language
1. Fire Head	One: will, desire, intent, direction, purpose, motivation. Zero dimensions. Unity.	Why Questions
2. Voice	NOT: logic, reflection, inversion, choice, free-will, speech. Division	Which Conjugations
3. Air Lungs	Two: duality, relationship. Distance, difference, division. Information, knowledge. Law, plan, path, design, rules, judgement. One dimension. Properties. Numbers.	How Adverbs

4. Heart	OR: measured-time, alternation, repetition, cycles, right-angles. (Addition / subtraction. Charge / discharge.)	When Time-Joiners
5. Water Stomach	Four, the sinewave, circles, cycles, seasons, waves. Action, motion, force, power, ability, work, people. Two dimensions. Vectors. Complex numbers.	Who Adjectives
6. Sex	AND: creation, (re)production, combination, mixing. "Earth in motion", so location. Multiplication.	Where Preposition
7. Earth Limbs	Seven / Eight. Physical matter, product, result, reward, material wealth. Three dimensions. Solid objects.	What Determiners + Pronouns

Fire: Will

Each successive level provides more detail and specificity. Fire is like one but adds detail.

Fire is "will", "desire", a form of consciousness. Fire explains that a fundamental aspect of "one" is "will". One has desires, and indeed must have to explain why the universe would be created at all.

Desire is the motivator, the driving force of all acts of creation. The tea won't get made if no one wants tea. All creation begins with "Fire". It tells us "why" things are.

Voice: Reflection

Duality is created by the "reflection of one". One is "reflected" or "inverted" into its logical opposite, "many". This super-concept of "reflection" is correlated with the principle of logic / reason, and the question "which" implying "this not that". It is the archetype of "choosing", the logical NOT and "division".

This isn't an object, it's a mechanism, a "verb". It has the power to "act on" the noun "one" and "transform" it into something else.

Air: Plan

Air is like Two, but more detailed. It is the super-concept of plans, designs, rules, and laws. A plan implies there is a right and wrong way of doing things. There is one right way to follow a plan, and infinitely many wrong ways.

Air corresponds to information that can be used for comparison. It tells us "how" things should be.

Heart: Alternation

The principle of alternation is the next mechanism / verb in the algorithm. It is associated with measured time, and cycles. It can be viewed as the principle of having two alternative paths to follow or alternation between two states.

It is by "alternating" the two halves of duality that the next conceptual object, and layer of the hierarchy, is created. It contains four sub-categories (as 2 x 2=4).

Water: Work

Four is a "duality of dualities". This level of noun-like concepts are two-dimensional, and allow the description of work / action. For example, forces in physics are represented as vectors which are two-dimensional numbers.

This is the level where "all the work gets done" (by enacting the Heart mechanism). The four sub-categories of this level match the traditional alchemical four elements, which is why their names are re-used here.

Note again that the system of categorisation is self-referential. The four elements categorise themselves as the "Water" level of the hierarchy. This identifies their role.

Sex: Mixing

Mixing, combination and sex are ways to describe this principle, the third and final mechanism. It is by mixing the four elements from above that we obtain the final three entities (Voice, Heart, Sex), and the full picture of seven principles in the final level.

Earth: The Product

This is the final level, and it contains these seven basic archetypes.

While Yin is the archetype of "matter" in general, Earth is the archetype of "the product", "wealth" and "value"; it is "things that matter", "that which is desired".

It is also the archetype of things that can be built on or used as a "foundation".

This archetype explains that the "foundation" of reality is not specifically physical matter, but "things that matter" in the most general sense.

"Earth" also contains ideas such as "fully formed", or "finished". It includes physical matter and also "data", which is "fully formed information" or "facts". It's a more specific version of Yin, and it explains that Yin is the "object of desire". "Yang desires Yin". As a rule:

"The creator desires the creation".

This list of seven super-concepts describes the "universal plan of creation", and we will explore how various natural phenomena correspond to it and can be better explained by it. It describes "what things look like on the outside", it's the basic blueprint for shape/form/matter. How things are made "on the inside" is a bit more involved, but this conveys the basic story.

These foundational archetypes are very generalised and broad. They are currently unrecognised by science as distinct entities, so we have no proper names for them. The labels used here are just suggestions.

Note that they correspond, in order, to the human body, as shown in the image below.

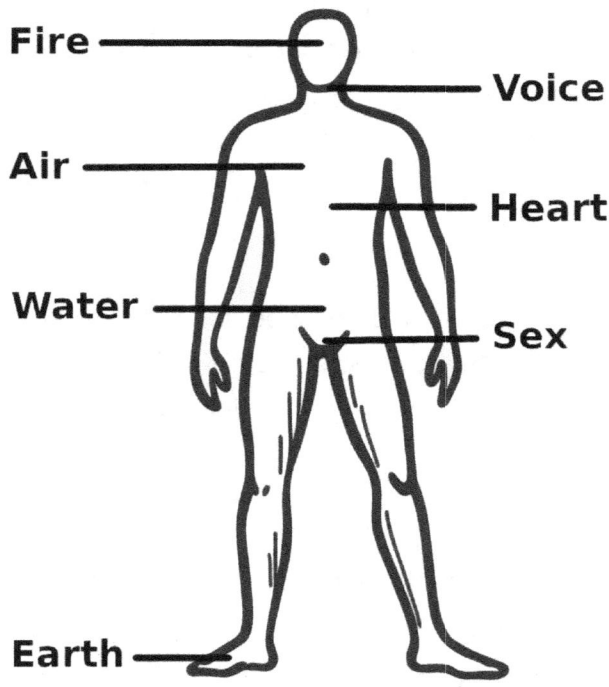

This ordered series of concepts is like a "template for everything", it's a single tool that can do every job, a language that can describe any conceivable thing. This diagram shows the ten basic archetypes and their relationships.

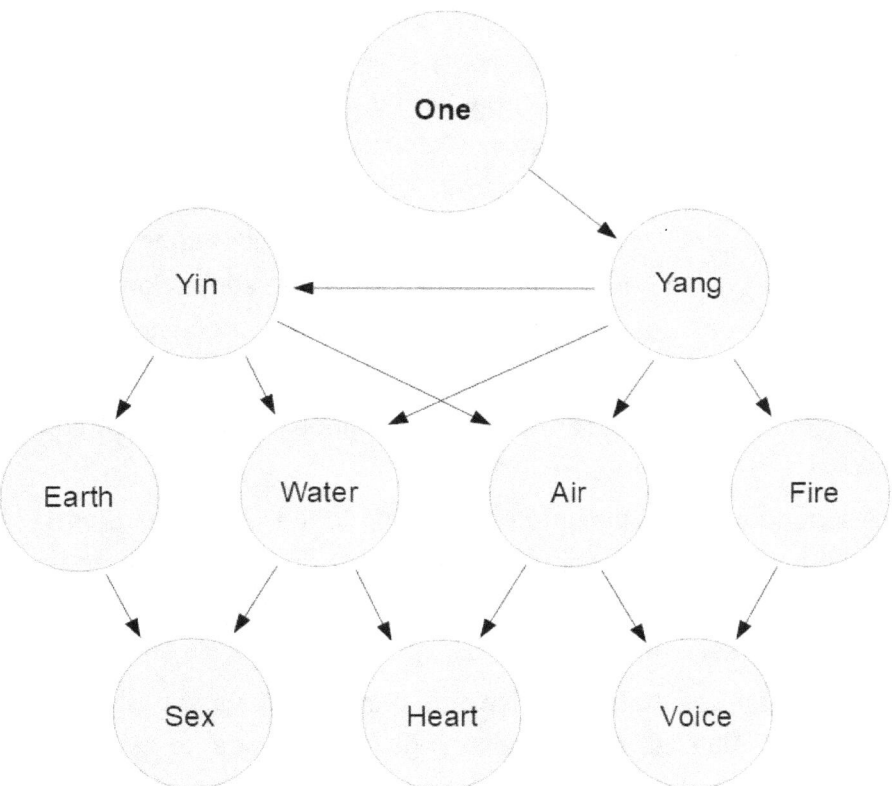

This section summarised the system; the rest of the book explains how it can be derived from basic observation and reason, with some discussion of the implications.

Some Definitions

The whole purpose of this book and the theory it describes is to define concepts. It's all about definitions, and there will be quite a lot of them. The book must attempt to describe the fundamental archetypes via as clear and unambiguous a path as possible.

How we define words is crucial to our understanding, misunderstandings are usually due to people having different interpretations of the same word. I'll use plain language as much as possible and avoid jargon and unnecessary technicality.

Some definitions in this book are from online dictionaries and search engines, some are my own.

I'll use quotes around words to indicate I'm referring to the word in its most general sense. *E.g.* when talking about "distance", the quotes emphasise the abstract nature of the concept of distance. Also, I'll use them to try to disambiguate where I feel it might be necessary. I apologise if it's annoying.

Although this theory argues that matter isn't the primary substance, I'll use the words "physical" and "material" to describe things which appear to be physical to us. While matter may not be the primary component of reality, it can still be considered "real", as real / illusion are relative concepts.

I will use the terms "reality" and "the universe" interchangeably.

universe
- the totality of known or supposed objects and phenomena throughout space; the cosmos

reality
- the totality of all things possessing actuality, existence, or essence
- that which exists objectively and in fact

This book argues from the idealist perspective: that the universe is made of ideas / information. Its opposite is materialism.

idealism
- the (philosophical) theory that the universe is made of ideas

materialism
- the theory that physical matter is the only reality and that everything,

including thought, feeling, mind, and will, can be explained in terms of matter and physical phenomena

The words "idealism" and "idealist" can be taken to mean "to idealise things" or "the belief that your ideals can be achieved", but these are not the definitions used here. This book only uses the word to refer to the philosophical theory.

The word "idea" just means anything that exists within a mind, so "idealism" could be defined as "the theory that reality exists within a universal mind".

idea
- something, such as a thought or conception, that is the product of mental activity
- any content of the mind, esp the conscious mind

Mentalism is essentially a synonym of idealism, and physicalism of materialism.

mentalism
- the doctrine that mind is the fundamental reality and that objects of knowledge exist only as aspects of the subject's consciousness

physicalism
- the doctrine that all that exists is ultimately physical

This theory aims to dispel "mysticism", so it needs a definition.

mystical
- "spiritually allegorical, pertaining to mysteries of faith," from Old French mistique "mysterious, full of mystery"

```
https://www.etymonline.com/word/mystical
```

Mystical implies "mystery", but that is "not-knowing". The problem with mysticism is it tends to correlate "not knowing" to "spiritual", which is a fallacy. It implies "mystery" is something "spiritual", and therefore does not need to be rectified. It effectively elevates "not knowing" to the status of "God", which is a paradox.

"Mysticism" in this sense conflicts with science. The only room in science for mysteries is as problems to be solved.

The etymological dictionary doesn't link "mystery" to "mist", but it seems possible these words are linked. A mystery is something that's "hidden in the mist".

Concepts

concept
- an idea, thought, notion

What a "concept" is may perhaps not immediately obvious. We use them all the time, we take them for granted, but what are they?

Concepts are something like "packets of information". They contain facts, logic, and references to other concepts. They are "thought forms", immaterial entities made of information.

Perhaps the best way to visualise a "concept" is as an online encyclopedia page. A concept is just like webpage containing a list of facts, opinions, images, videos, and references to other related concepts.

I suggest that the world-wide-web architecture (that was enabled by the invention of HTML by Tim Berners-Lee) is a fairly accurate external representation of how thoughts are stored inside our minds. The internet has been so successful because it mimics the structure of the human mind so well.

(Note: the word "concept" is pronounced "con-sept", which would break down into con- {with} and sept {seven}. This isn't believed to be the etymology of the word, but it is interesting to note.)

Abstract Concepts

To describe anything, we must use abstract concepts (or "abstract objects"). All matter has properties like size, mass, shape, location, and may have others like colour, smell, texture *etc.* All properties are abstract concepts.

The idea of "size" is a feature of nature. It's not a material object, it's a "quality" material objects have. It's an idea necessary to describe matter. Abstract concepts like this are the building blocks of our thought processes.

abstract
- relating to or involving general ideas or qualities rather than specific people, objects, or actions
- to make a summary of
- literally to "bring out", as in to separate or extract the essence of something

Just as an abstract of a scientific paper is a summary of it, it's also an "extract". It is a "distillation" of the "spirit" of the observed phenomenon.

abstract concept
- a principle, or set of principles, abstracted from observation of matter /

nature, or from other abstract concepts
- a distillation of principles from a greater body of information

principle
- essential quality; law, or rule

We refer to (some/all?) abstract-concepts as "abstract nouns".

"Abstract nouns represent intangible ideas things you can't perceive with the five main senses. Words like love, time, beauty, and science are all abstract nouns because you can't touch them or see them."

https://www.grammarly.com/blog/abstract-nouns/

The word 'abstract' can be taken to mean "difficult to understand" or "insufficiently factual", but this is not the meaning implied here.

The "Conceptual Framework"

The position of materialism is that abstract concepts only exist in the brains of people and not "out there" in the universe. In the idealist view there is a "world of concepts / archetypes", a "spirit-world", a non-material universe populated by the rules and principles that the material world is contingent upon. These are sometimes called "platonic objects".

Numbers are "spirits". The number "one" is "just an idea". You can have examples of it in matter, but abstract concepts aren't made of matter they're made of "spirit".

In idealism such non-material objects are the building blocks of reality. Idealism generally implies that things must exist as concepts before they can be manifest into physical reality. The idea / thought must come before the act. (Just as a virtual reality must be programmed before it can be run.)

There are quite a few terms we could use to describe the system under discussion including: the world of concepts *conceptual space* conceptual framework / the UP. I'll mainly use the terms "conceptual framework" and "UP", but I'll use others too for variety.

conceptual framework (my definition)
- the hierarchy of fundamental ideas which exists in the "universal mind"

This theory also describes an ontology of abstract concepts / universals.

ontology
- a rigorous and exhaustive organization of some knowledge domain that

is usually hierarchical and contains all the relevant entities and their relations

The framework is not just a passive list of categories, it contains instructions as well. It's an "algorithm", a "program", a "story", a "template", and a "recipe".

algorithm
- a set of rules for solving a problem in a finite number of steps

story
- an account or recital of an event or a series of events

template
- anything that determines or serves as a pattern; a model

recipe
- a set of instructions for making something

Words Are Labels

word
- a unit of language, consisting of one or more spoken sounds or their written representation, that functions as a principal carrier of meaning

A word is an arbitrary* label, used for communication, which represents something else. The word "dog" is not the same as the idea of dog, which is not the same as an actual dog. Words are labels which represent or refer to ideas or material things, they are not the idea itself.

actual dogs > idea of dog > word "dog"

The "idea of dog" can be represented by any words / combination of sounds, as long as everyone (local) agrees on it. A person who speaks Spanish has the same basic idea of dog as one who speaks English, but they use different words to describe that idea. The fact that the "idea of dog" is held in common by people, no matter what language they speak, is what allows us to translate between languages.

The extent to which different concepts are universal is a complex issue. There are some which do seem to be truly universal among all cultures and languages, and others that are less so.

Note, natural semantic metalanguage theory has attempted to identify the most elementary linguistic concepts that have the same translation in every language. "The natural semantic metalanguage (NSM) is a linguistic theory that reduces lexicons down to a set of semantic primitives."

```
https://en.wikipedia.org/wiki/Natural_semantic_metalanguage
```

* It's generally assumed that languages are arbitrary because there are so many of them, and they differ so much, but the UP indicates there might be a "right answer" to the question of language.

Universals

"Universal" is a generalised term for all types of abstract concepts / objects, so called because they exist universally, at all times and locations, in the same (non-physical) form. There's currently no standardised list of categories of universals, this theory attempts to provide that.

universals
- term applied to general or abstract objects such as concepts, qualities, relations, and numbers, as opposed to particular objects.

Materialism views universals as existing only in the human mind, idealism views them as having independent existence.

"The exact nature of a universal deeply concerned thinkers in the Middle Ages. The extreme realists, following Plato, maintained that universals exist independently of both the human mind and particular things.
In nominalism universals are considered arbitrary constructions of the human mind.
In conceptualism universals exist only in the mind, as concepts, but they are not arbitrary, as they reflect similarities among particular things. "

```
https://encyclopedia2.thefreedictionary.com/universals
```

The UP describes an essentially "realist" philosophy, that fundamental universals exist independently of human minds, but it also says there's a spectrum of "reality". The duality of real *illusion is a sliding scale, not an on*off switch.

The foundation of reality is universal *general, but it tends towards being local* specific as it descends through the levels and grows in complexity. The UP is the most real set of universals, everything derived from them is less real / more illusory.

The UP says it is possible for human minds to discover new things and invent novel ideas. We can create new lower order / mixture universals which did not previously exist in the universal mind, and that is our purpose. The idea of "cheese sandwich", for example, is a "relatively universal" archetype, invented by humans.

Life exists to discover and describe the fundamental archetypes from an independent perspective, and to invent new mixtures of them.

Information

A universe can, in theory, be made of information.

information
- facts, data, or instructions in any medium or form

Information creates a kind of form or shape within a mind / memory. It informs, it "makes a form inside".

inform (etymology)
- from in- "into" (from PIE root *en "in") + formare "to form, shape," from forma "form"

https://www.etymonline.com/word/information

Simplistically, when you see a new thing, you remember what it looks like; you form an image of it in your memory. The next time you see an object, you can compare it to your collection of stored images to see if it's similar.

Computers are the same, information is stored as physical shapes / forms on some kind of medium, usually as magnetic or electric fields. While information is an intangible thing it has a kind of form, or at least it can be represented by forms.

Information is always represented via "shapes", but the shapes themselves are not the information, they just convey it. For information to have "meaning", it must be "interpreted" by a consciousness.

Knowledge

I'd like to draw the following distinction between knowledge and belief.

knowledge
- information we can demonstrate to be true (at least to ourselves)

belief
- information we consider to be true but lack evidence for

Theory of knowledge, how we know things, is "epistemology".

epistemology ... is the branch of philosophy concerned with knowledge. Epistemologists study the nature, origin, and scope of knowledge

https://en.wikipedia.org/wiki/Epistemology

It's a complex subject, but the basic approach taken here is that we can only have "knowledge" of things we have experienced directly. The strictest application of the rule would be:

- I know I exist, and that ideas exist.
- I believe the physical world exists, but I can't prove it.

We experience concepts directly, by seeing them with the "mind's eye, so they are the most "knowable" things. We experience the universe indirectly, via our senses, and they might not be telling us the truth. We might be in "the matrix".

Strictly, I can only "believe" the physical universe exists as I perceive it. However, it's reasonable to extend "knowledge" to things we experience via the physical senses, within the context of this reality. Our minds and physical experiences are the most "real" things. So, we'll define knowledge as including the physical world.

knowledge
- personal experience, both of mental and physical phenomena.

If we have personally experienced (or understood) something, only then can we say we "know" it. Anything else is a "belief". This may seem an overly strict definition, but it has the virtue of being relatively unambiguous if we need to make decisions based on it. It is the same standard as we would find in a court of law where only witness testimony is considered evidence.

However, outside of a decision-making context, this is a sliding scale. Generally, we start out with beliefs and work our way toward knowledge. Science requires multiple evidences to support a theory. As a general principle though, it is "observation" that allows belief to become knowledge.

Only (personal) observation can provide knowledge.

Meaning

We need a working definition of "meaning".

meaning
- the end, purpose, or significance of something.

If we know the meaning of something, we understand it.

understand
- To become aware of the nature and significance of; know or comprehend

The word "signify" is a synonym of "symbolise", so we could say that meaning is determined by what a thing symbolises.

signify
- to denote, to mean
- to be a sign or indication of; suggest or imply
indicate, show, mean, imply, convey, symbolise, denote, portend
from Latin significare, from signum a sign, mark + facere to make

To symbolise is to "represent", which means to act on behalf of another. It's a relationship with another concept. A thing's significance is determined by what other ideas it links to. A full discussion on the concept of meaning could be a long one, but our working definition will be that it is a measure of the relationships between ideas, *i.e.* "meaning" comes from the connections (comparisons) between concepts.

This links to the observation we made earlier that all descriptions are comparisons.

Things only have a describable meaning via contrast with other things. To describe a thing, it must be compared to something else. Ultimately all ideas share some relationship with each other, just as all physical things do. Everything is related to everything else to some degree or other, and "meaning" is found in those relationships.

A dictionary is a book which describes the meaning of words by linking them to other words which have different meanings. It's not possible to define a word with a synonym.

Categorisation and Sets

The system this theory describes is a hierarchy of (something like) categories or sets.

category - a specifically defined division in a system of classification; a class

Some principles of set-theory could be applied here, but set theory is a bit over-powered for what we need. It's worth a mention though. The principles of set theory are used a lot in databases.

Set theory is the branch of mathematical logic that studies sets, which can be informally described as collections of objects.

https://en.wikipedia.org/wiki/Set_theory

Sets are collections, lists, categories, groups. These are all just different names for a one-to-many relationship. The underlying principle we're interested in is the one-to-many relationship.

The basic idea of set theory is quite simple, and we use them intuitively in everyday life. A shopping list is a "set of things I need to buy". A one-to-many relationship is like a "container"; it's the idea that one thing can "contain" a number of other things.

Sets are containers; they describe things by containing them. We can define a set called "all dogs" as including all dogs in the world. The "set of all dogs" is just an idea, it can't physically contain anything, but it can contain ideas / labels. The "set of all dogs" is a lot like the word "dog".

The Universal Set

In the system of classification being described, we need some kind of top-level category; a single container to hold all the other ideas. In set theory this is the "universal set", but things can get muddled because sets may be defined as "containing themselves", depending on which version of the theory is being used.

In set theory, a universal set is a set which contains all objects, including itself.

`https://en.wikipedia.org/wiki/Universal_set`

In the system being described no set or category can contain itself. It's a strict hierarchy of categories, so this paradox doesn't apply. I'd argue that nothing can "contain itself" as that confuses container and contents, which is a category error.

A Theory of Categories

One of the most profound questions that has exercised the minds of philosophers throughout history is the question of categorisation. How should we categorise phenomena?

What are the most fundamental categories of things?

If we could categorise everything into a minimum number of "types" of thing, what would those categories be?

In ontology, the theory of categories concerns itself with the categories of being: the highest genera or kinds of entities. Various systems of categories have been proposed, they often include categories for substances, properties, relations, states of affairs or events.

`https://en.wikipedia.org/wiki/Theory_of_Categories`

It's hard to overstate the importance of this endeavour. Categorisation is equivalent to understanding. If our system of categories is faulty in some way, then it will be impossible for us to understand things properly.

We understand things by grouping them into categories to which properties are assigned. If we miscategorise things that means we have assigned the wrong properties to them. If we do not have an objective system of categorisation then we will inevitably miscategorise things, thus fail to understand them.

In the absence of an absolute, objective, correct system of categorisation, we can have no idea just how wrong our understanding might be. There is no way to measure accurately if all rulers are faulty. We might think we're close to the truth but be a million miles away.

Categorisation is the foundation of understanding. If we can't categorise things, we don't understand them. A philosophical "theory of categories" is like a conceptual "theory of everything".

There have been many (wildly different) schemes proposed by famous thinkers from Aristotle onwards that attempt to answer this question, but there has been no theory proposed that seems to be truly universal. The UP, if found to be correct, would be the definitive answer to this question.

A Theory of Everything

What is a theory of everything (TOE)?

The predominant theory of reality in the world today is materialism, and this is reflected in Wikipedia's definition where it's defined as a theory concerned with matter only.

"Theory Of Everything is a hypothetical, singular, all-encompassing, coherent theoretical framework of physics that fully explains and links together all physical aspects of the universe"

`https://en.wikipedia.org/wiki/Theory_of_everything`

When physicists talk of a TOE, they mean a way to unite the systems of general relativity which deals with gravity, with the standard model which deals with the other three forces. In this book the term will be used in a more general way, this theory isn't limited to only explaining physical phenomena.

theory of everything (my definition)
- a system of knowledge from which all natural phenomena can (conceivably) be derived.

The theory we're looking for isn't a unification of the two branches of physics but of all concepts and phenomena.

Both material science and the UP aim to explain how the universe is constructed and what it's made from, so it's reasonable to use the term in both contexts. However, physics takes mathematics for granted. A theory that can't account for the existence and structure of logic and mathematics can't be a complete TOE.

Chapter 2

Duality

The Origin of Information

If we're looking for structure in our conceptual framework there's one obvious place to start. All information is founded on the principle of duality, just as information-technology is founded on binary. The existence of duality (within the world of ideas) is an objective fact.

There is no doubt duality exists as a rule / principle of thought. Most other concepts depend on it, including logic and maths. It's a small piece of solid ground we can start to build on. Duality is the foundation of all conceivable thought or computation. All fundamental properties (like "distance") are conceptual dualities. In this section I want to highlight the foundational nature of this principle.

Before I began this investigation, I was quite sure I understood the number two. It seemed to be such a simple idea. It didn't seem like there was that much to know about it. I couldn't have been more wrong. The concept of duality is a bit like a doorway: on the outside it doesn't seem very large or interesting, but if you look closely it opens up, and there's a lot more to see inside than you might think.

Understanding the nature of duality is the key that unlocks all the answers we're looking for. It's like a "gateway to knowledge".

Oppose / complement

Dualities are fundamental abstract concepts / universals containing two opposing or complementary possibilities.

duality - the state or quality of being two or in two parts; dichotomy

dichotomy - division into two parts or classifications, esp when they are sharply distinguished or opposed

polarity - the possession or manifestation of two opposing attributes, tendencies, or principles

How many distinct components is the universe made of?

monism - the view that there is only one basic substance or principle as the ground of reality or that reality consists of a single element

dualism - the view that the world consists of or is explicable as two fundamental entities, such as mind and matter

pluralism - the doctrine that reality is composed of many ultimate substances

This theory suggests all the above views have some validity depending on perspective, but level two is where things start to be explained. It's the beginning of knowledge because we can only know things by comparison with other things.

Duality As Relationship

Duality is closely related to the concept of "relationship". All fundamental properties are dualities and describe relationships (high/low, big/small, hot/cold, etc.)

relationship - the condition or fact of being related; connection or association

Everything that exists has some relation to everything else and can only be described by its relation to other things. It's impossible to define any idea without relationships. This means there can be no information without duality. Duality comes before information. It must exist before any other ideas can be defined, so it must be at or near the top of the hierarchy of ideas.

If information is "made of duality" which is "relationship", and the universe is made of information, then:

The universe is made of relationship(s).

The principle of duality is probably the Alchemical "prima materia". The "first and finest form of matter".

https://en.wikipedia.org/wiki/Prima_materia

Duality In Language

The importance of the principle of duality in all forms of language, logic, and cognition can't be overstated. It underlies and enables all other concepts and even knowledge itself. Duality is like a "parent" concept from which other ideas can come.

From the article "Duality in Logic and Language" on the Internet Encyclopedia of Philosophy:

Duality phenomena occur in nearly all mathematically formalized disciplines, such as algebra, geometry, logic, and natural language semantics.

`https://iep.utm.edu/duality-in-logic-and-language/`

The article continues.

Duality phenomena of this kind are highly important. First of all, since they occur in formal as well as natural languages, they provide an interesting perspective on the interface between logic and linguistics. Furthermore, because of their ubiquity across natural languages, it has been suggested that duality is a semantic universal, which can be of great heuristic value.

A "semantic universal", is a concept which has meaning in all cultures.

semantic
- of or relating to meaning, especially meaning in language

Duality underlies everything that is conceptually meaningful. It's impossible to imagine any form of existence that doesn't rely on duality. The most basic form of reality must include the "observer" and the "observed", which is a complementary duality. We should consider the concept as more fundamental than any other.

Components and Connections

The basic question this book attempts to answer is: how is this reality constructed?

One thing materialists and idealists would probably agree on is that reality is made of a finite number of relatively simple components which combine to produce more complex things. They may disagree about what those basic parts are and how they interact, but the concept of a construction necessarily contains the duality: components and connections.

construct
- (v) To form by assembling or combining parts; build.
- (n) Something formed or constructed from parts. Something formulated or built systematically.

component
- a constituent part; element; ingredient.

connect
- To join or fasten together

The concept of "construction" says reality is made of exactly two things (or types of thing).
1. Components *parts. A noun* object / passive part.
2. Connections / arrangement. A verb *action* active part.

When a bricklayer builds a wall, two things are required.
- Bricks. The components of the wall.
- Mortar. To connect the components.

In the materialist view the components of reality would be "particles", connected by "fields" or "forces". So, particles are the passive bricks and fields are the active mortar. In idealism, the bricks are "platonic objects" and the connections are their relationships.

Note that the bricks in a wall are discrete and can be counted. There are many of them and each is separate from all others. The mortar on the other hand is continuous, there's only one piece of mortar in a wall, it's all joined together. This embodies the one-to-many relationship we will find in all dualities.

Many discrete (Yin) components, one continuous (Yang) connector.

Distance

Distance, measured in meters (m), is a fundamental unit in physics. It's an abstract concept we're all familiar with, and it's a duality / relationship. Distance is a one-dimensional relationship.

Everything that exists has a distance from every other thing. Distance is the same thing as space, and time can also be thought of (and depicted) as a distance. Einsteinian "spacetime" is made of "distances".

distance - The extent of space between two objects or places; an intervening space. - a separation or remoteness in relationship; disparity
The word's origin is:

*distance (etymology) - "stand apart," from dis- "apart, off" (see dis-) + stare "to stand," from PIE root *sta- "to stand, make or be firm."*

https://www.etymonline.com/search?q=distance

At its most basic, distance is a degree (or measurement) of separation. For there to be a distance we can measure, there must be two things to measure between. For there to be two (or more) things, (some form of) distance must exist.

Conceptually, the ability for things to be separate must exist before there can be two things. If there was only one thing in existence, there could be no distance or separation. "Distance" is closely related to "difference".

difference - dissimilarity, unlikeness, divergence, variation, distinction, discrepancy

Any difference between two things can be treated (conceptually) as a distance. Distance *separation* difference is a fundamental principle of reality. If reality is constructed from ideas, then distance must be one of the very first ideas that must be defined because almost every other concept depends on it somehow. The existence of nature, or you and I, depends on the principle that things are separate, apart, and different.

We can talk of emotional distance in a relationship, or intellectual distance between theories. "Distance" can apply to many conceptual relationships. Synonyms of the word can refer to physical properties like:

space, length, extent, range, stretch, gap, interval, separation, span, width

Distance can refer to personal *emotional* other types of difference:

aloofness, detachment, indifference, coldness, remoteness, disagreement

Dimension

When we measure a distance, we have obtained a "dimension". There are three dimensions of space, but anything we can measure can also be considered a dimension.

dimension
- a measurement of the size of something in a particular direction, such as the length, width, height, or diameter
- (physics) A physical property, such as mass, distance, time, or a combination thereof, regarded as a fundamental measure of a physical quantity.

dimension (etymology)
- from PIE root *me- (2) "to measure."

https://www.etymonline.com/word/dimension

Although the etymological dictionaries don't go this route, the word could perhaps also be broken down as:

dimension
- di (two) + mens (mind) + ion (process). A "process of two minds",

A dimension is a duality, its two options can be considered as two "minds" (sets of concepts, "near/far"), and an extent or distance must be created by a process of some sort.

Fundamental Properties

fundamental
- of or relating to the foundation or base; elementary
- something that is an essential or necessary part of a system or object.

The concept of distance is fundamental. It's a necessary part of reality. It contains two sets of opposing ideas like near *far, close* distant. Near and far are gross generalisations and they only have meaning in relation to each other, they are relative not absolute. This is the same for all descriptive dualities, they're all relative, they only gain meaning in comparison with each other.

Many properties are not dualities, such as flavour, nationality, shape, or colour, but all the fundamental properties that we use in physics, like mass, charge, voltage, current, temperature *etc.* are.

All fundamental properties are dualities (relationships).

The theory predicts:

All fundamental natural phenomena consist of two parts or come in two forms.

Anything that is a fundamental feature of nature must be a duality, a one-dimensional relationship.

Two Types of Duality

There are two types of dualities: opposites and complements.

Opposites oppose each other; they act as a counterbalance or contrast to each other, whereas complements complete or match each other. Opposites conflict or "compete" like "enemies" and are mutually exclusive, complements match or "cooperate" like "friends" and are "inclusive".

oppose
- to present in counterbalance or contrast
- to be or act in contention or conflict with
conflicting, different, opposed, contrasting, opposite, differing, contrary, contradictory, incompatible

Some examples of opposing dualities:
- Distance - near / far
- Weight - heavy / light
- Light - dim / bright
- Temperature - cold / hot
- Velocity - fast / slow

complement
- to complete; form a complement to
matching, companion, corresponding, compatible, reciprocal, interrelating, interdependent, harmonizing

Some examples of complementary dualities:
- A couple - husband / wife
- Teaching - teacher / pupil
- Ancestry - parent / child
- Driving - driver *car, or captain* ship etc...
- Trade - seller / buyer

Sadly, this isn't suggesting that "a couple" in real life must always be in harmony, it's saying the concept of "a couple" relies on the concept of harmony / complementarity.

Discrete / Continuous

Information can be represented in either a discrete or continuous form. This relationship is fundamental and extremely useful, we will come back to it many times. Opposing dualities are expressed as a range, a spectrum of values, but complementary ones are either on or off. There are infinite possible points between any two opposites, it's a continuum, but you can only either be a driver or not be one. It's digital, on/off, true/false.

continuous

- uninterrupted in time, sequence, substance, or extent

constant, continued, extended, prolonged, unbroken, uninterrupted, unceasing

discrete

- separate or distinct in form or concept

- consisting of unconnected distinct parts

separate, individual, distinct, detached, disconnected, unattached, discontinuous

In the context of information, continuous data includes "analogue" data and the set of real numbers. Discrete data includes "digital" data and the set of integers.

analog / analogue

- of, relating to, or being a device in which data or a signal is represented by continuously variable, measurable, physical quantities, such as length, width, voltage, or pressure

digital

- expressed in discrete numerical form, especially for use by a computer or other electronic device.

When building a wall, the bricks are the discrete components, and the mortar is the continuous component. There are many individual bricks in a wall, but the mortar is all joined together as a single thing.

In a song, the musical notes are many and discrete, but the composition is continuous and singular. In a book, the words used are discrete, but the message they convey is continuous. There are limited number of musical notes and words, but an infinite amount of possible music and writing because a finite number of components can be connected in infinite ways.

Here's a comparison between discrete and continuous data, they mirror each other.

Feature	Discrete / Digital Data	Continuous / Analog Data
Number Type	Integer Numbers, Finite, Limited	Real Numbers, Infinite, Unlimited
How To Quantify	Countable	Measurable
Type of Graph	Bar-Graph	Line-Graph

Accuracy / Resolution	Limited internally by design, number of bits	Limited externally by accuracy of measurement
Exactitude	Can give exact results	Cannot give exact results
Processing	Multiple steps	Single step

Note that we can also use the word "analogue" in the following sense.

analogue - something which is analogous (similar, corresponding) to something else

Perhaps the word has both meanings because all the "parts" of a continuum are the same whereas all the parts of a list of discrete items are different.

A Spectrum of Possibilities

spectrum - a continuous sequence or range

continuum - a continuous extent, succession, or whole, no part of which can be distinguished from neighbouring parts except by arbitrary division. e.g. the set of real numbers

All properties that are opposing dualities can be thought of as a one-dimensional (1D) object like a string; they describe a conceptual line between two extremes like "cold and hot" or " long and short". *I.e.* a thermometer, or a ruler.

Much of science relies on the premise that representing the fundamental aspects of reality as 1D relationships is a valid approach. It works out mathematically, and seems to be an accurate representation of reality. The UP supports this approach.

A Spectrum of Languages

As mentioned above, information can be conveyed in continuous or discrete form, as analogue or digital data. Conveying information is "language".

We should note that the duality discrete *continuous is not complementary, so must be an opposing duality. Discrete* continuous is thus a continuum. Information can be conveyed in a range of ways between fully discrete and fully continuous.

There is a spectrum of possibilities between the two options, and all information-transfer methods (languages) must lie somewhere on it.

Ultimately, all information is conveyed in "packets" which can contain variable amounts of information.

Binary is the most "discrete" form of language.

It has the smallest possible amount of data, just 1 or 0, in each "packet" or "bit".

The UP is the most "continuous" form of language.

It has the largest possible amount of data contained in each "packet".

Human languages are somewhere in between.

Able to convey highly variable information with "words", where each word is like a "bit".

"String Theory"

A short diversion: in a way, the UP implies a type of "string theory". Properties are 1D objects, and that is the essence of a "string". It suggests that reality is fundamentally made of dualities, which are like 1D strings made of information, and these are somehow "woven together" to create matter. These strings are the "properties" the object has, maybe qualities like colour or weight, temperature or charge.

One-dimensional objects are "everywhere in the universe at all times" because they can't have a location. The same applies to 2D objects. Things must have three dimensions to be located in 3D space.

How exactly these "threads" might be "woven" into reality is a topic for another day, I just thought the idea was worth mentioning.

One / Many

At the level of duality there can only be two numbers. In binary, they would be one and zero. Alternatively, we could have one and "many" as our number line. All hierarchies involve one-to-many relationships, and all information systems have hierarchies.

- One company has many employees.
- One employee has many tasks to do.
- One task consists of many sub-tasks.

The one-to-many relationship and its variants are of particular interest to database designers as they dictate how the data is naturally structured, and hence how it should be represented in a database. It is going to be one of the main tools we will use to deconstruct reality and find out what's underneath.

The possible list of relationships in a family are:

- One to one (e.g. father to mother).
- One to many (e.g. mother to children).
- Many to one (e.g. children to mother).
- Many to many (e.g. siblings to each other).

If two things have a one-to-many relationship, then they exist at different "levels" or "generations" in a hierarchy. One always comes before many. We'll be seeing this principle a lot.

Deriving Numbers from Duality

Can we logically derive all numbers and mathematics solely from the concept of duality?

derive
- to arrive at by reasoning; deduce or infer
deduction: facts are determined by combining existing factual statements
induction: facts are determined by repeated observation

Arguably, "two" contains within itself the potential for all the numbers, or at least all the positive integers. It contains a kind of logical blueprint or formula for making numbers. All concepts contain their own facts and logic, and the concept of "two" is like an algorithm. If its own internal logic is applied, all (integer) numbers can be deduced from it quite easily.

If the number two is made of two ones, then:

- Numbers can form a progression.
- Numbers can be combined to create greater values (addition).
- Numbers can be separated to create smaller values (subtraction).

"Two" contains enough data within itself to easily derive all positive integers, and the concepts of addition and subtraction, so it leads to zero and negative numbers too.

The duality of multiplication / division is sort-of implied by the fact that two can be divided into two ones, but these ideas aren't very useful when the only numbers that exist are 1 and 2.

We could go on to consider ways that other types of numbers (fractions, reals etc.) might be logically arrived at, but it would be better to revisit the idea once the full groundwork of the theory has been laid.

Measurement Creates Quantities

Taking measurements is the "bread and butter" of science. Data is the foundation of knowledge, without it, we couldn't even have theories. Measurement is closely related to the concepts of observation, perception, experience, comparison, and judgement.

measure
- (noun) a reference standard used for the quantitative comparison of properties
- (verb) to bring into comparison, quantify, determine, judge, compare, evaluate

A measurement in science is an "observation".

observe
- to see; perceive; notice, to watch attentively

observation
- the act of observing or the state of being observed
- the facts learned from observing

Measurement is observation, it bridges the gap between observer and observed. It is the verb "to look", enquiry, investigation, and comparison. The act of measurement is a deep archetype; it represents the "desire to know". It's the verb which joins the nouns of consciousness and experience, the inner and the outer worlds. Note that all sensory perceptions are measurements, eyes measure light, ears measure sound, and so on.

Measurement also bridges the gap between the qualitative and the quantitative realms, which is another important duality to consider.

qualitative
- relating to the quality of a thing, measurable but not countable

quantitative
- relating to numbers or amounts, countable, measured, a measurement

Before you measure something, like a parcel, it only has a qualitative size, such as "medium sized", or "small". When you measure it, you obtain a quantitative size, like "20cm".

Measurement is the mechanism that creates quantities from qualities. It is how things can be better "known" or "described" (written down), and it works by comparison. We started out with a personal one-dimensional qualitative description like "small", and moved to a shareable, two-dimensional one, "twenty centimetres". (The number and units are both dimensions, we must specify both to convey the data.)

Qualities are one-dimensional. Quantities are (at least) two-dimensional.

Both the act of measurement itself, and the result it produces are comparisons, *i.e.* "twenty centimetres" is a comparison of the parcel with the unit "centimetre".

Measurements Are Relative

All measurements are comparisons with other measurements, such as "1/2 a mile" or "5 seconds". While qualitative statements have only one part like "near" or "far", quantitative measurements have two parts, a quantity and a quality, a number and "units". But note that all units are made of another measurement (e.g. "a mile"). They're more like a quality than a quantity, but they are still, at heart, just another measurement.

There are no absolute measurement units (that are easily determinable). A "mile" is a relatively arbitrary distance-unit against which other distances can be compared. The creators of our system of units have tried to relate them to real phenomena, to give them some grounding in nature, but there is no apparent or easily accessible absolute universal standard. (The Planck units might be.)

For example, we record time in a day as 24 hours x 60 minutes x 60 seconds, it gives us a total of 86,400 seconds per day. You would be forgiven for thinking it's just an arbitrary convention, but each time-second is equal to 15 arc-seconds of rotation of the Earth. It's not an arbitrary duration for us, but in most other places in the universe it would be. It's "relatively arbitrary" or "relatively absolute".

Absolute / Relative

In the process of measuring, we go from qualities to quantities, and from "absolute" to "relative" which is another important duality. The word "absolute" has several meanings, but here we're primarily interested in the context of it being half of the duality "absolute / relative".

absolute
- not relative to anything else, and/or used as a reference point for other things

relative
- something which is defined by its relation to something else

As a simple example, imagine a X/Y graph of some data. The origin of a graph (the 0,0 point) is the absolute, and any points drawn on the graph are relative to that. There's one origin, but many points on a graph. The origin comes first, the data afterwards.

When you see a scene with your eyes, the raw data your brain receives is the absolute, and the meaning you derive from it is relative. The scene is not relative to anything else but all the things you perceive within it are relative to each other, and to your knowledge. There's one set of input data, but many possible interpretations.

The Ruler

When we measure a parcel, the ruler is the "absolute standard", and the measurement is relative to it.

Measurements have rulers, and societies have rulers too.

ruler
- a straight-edged strip, as of wood or metal, for drawing straight lines and measuring lengths
- one, such as a monarch or dictator, that rules or governs

English contains many words with multiple meanings, and they often make interesting connections between seemingly disparate concepts. Rulers take measures. Government officials say they are "taking measures". Rulers provide the "law", the standard by which things should be judged. A "ruler" may judge things in units of "inches" or "innocence".

The archetype of measurement is like the "king" or "judge" of all concepts as it is the active process of "obtaining truth". It's the "judgement" which produces ","knowledge" which can be communicated to other minds (is "shareable").

(Note, the word root "reg-" as in "regulate", "regnum", "regis" also links "rule" and "royal". Also, the words "royal" and "real" appear linked via the underlying meaning "of God", "re-al / roi-al".)

The ruler represents the concept of having a standard, a "law-giver".

Division

The measurement archetype contains (or references) several other concepts. Perhaps the most important one is "division". The act of measurement divides the ruler into two parts.

divide
- To separate into parts, sections, groups, or branches
separate, part, split, cut (up), sever, partition, shear, segregate, cleave, subdivide

Measurement divides true from false, near from far, observer from observed, *etc*. We will see later how all information is created via the division of continua into discrete parts, so "division" is a key part of the process of obtaining information.

("Division" correlates with the Voice principle, "reflection", reasoning and discernment, the Boolean "NOT", etc)

When we make the measurement with a ruler there have three parts, one is active, the other two are passive. The measurement point (X) is the active principle, and the parts (A) and (B) are the passive result of where it's placed.

We will see this same structure, theme, or "motif" in many other places (as this is a representation of the One / Yang / Yin relationship).

```
          0 . 1 . 2 . 3 . 4 . 5 . 6 . 7 . 8 . 9 . 10 . 11 . 12
          |--------------------------------(X)------------------
          ^           (A) 8 units           ^    (B) Undefined   -->
```

The three parts are:
- (A) the length of ruler from the origin, 0, to the end of the object: *E.g.* "8cm"
- (X) the active measurement point. X "marks the spot".
- (B) the length of the ruler going away from the object.

Part (A) of the ruler is defined, we know how long it is.

Part (B) of the ruler is undefined, we don't know (or care) how long it is. The physical ruler has an end, but we could be taking the measurement with any sufficiently long ruler, so this part is only bounded at one end. We could say that (B) is infinitely long because it has no endpoint, whereas (A) has two ends. Its length is finite and known.

In contrast, (X), the measurement point has no size, no extent. A point has no dimension, it's just an idea, not a material object.

One is (X), Yin is (A), Yang is (B)

The Point

Between Earth and Sky

Yin is the Earth, Yang is the Sky, You are One.

When we stand on the earth and look around, the earth is like part (A) of the ruler. It's known, we can see where things begin and end, and we can measure them. The sky is like part (B), it's only bounded at one end. We can see where the sky begins (at ground level), but not where it ends, it goes on "forever".

Between Past and Future

We live in the present, which is (X). The past is like part (A) of the ruler, known, bounded at two ends (birth, now). The future is like part (B), only bounded at one end (now).

Life exists in between Yin and Yang. In between earth and sky, past and future, we find life, humans, consciousness. So, we are (X) the measurement point, consciousness.

Consciousness

The point (X) is an immaterial thing with zero size, but finding it is the method / algorithm used to solve the problem of discovering a distance or making a judgement. When we take a measurement, we measure up to "the point", and that's the point / purpose. Finding the point is the point.

point
- A dimensionless geometric object having no properties except location
- an objective or purpose to be reached or achieved

To take a measurement we must place the intangible, dimensionless point on the dimensioned measuring device by an act of conscious will. The (X) is located by "desire", "intent", "purpose", and it is always "made of consciousness".

Humans (strictly "individuated consciousnesses") are "the point" of creation, our purpose is to measure it from an independent perspective, objectively.

From Quality to Quantity

Materialism's view is that our (qualitative) perception of the world is generated from quantitative signals in the brain. It believes qualities are created from quantities, *i.e.* "spirit comes from matter".

From the idealist perspective the world is manifest from quality to quantity, "matter comes from spirit", and it is the act of measurement / observation which does that. It is the (only) process by which "spirit becomes matter" or a quality becomes a quantity.

A popular interpretation of quantum mechanics says that the world doesn't exist in solid form until it's observed. It says the world exists only as "probabilities" until an observation "collapses the wave function".

In quantum mechanics, wave function collapse occurs ... due to interaction with the external world. This interaction is called an observation...

https://en.wikipedia.org/wiki/Wave_function_collapse

The UP suggests that the reason physicists see the quantum world in this way may be due to the constraints of the conceptual framework, and there is a simpler interpretation.

In this view of QM, it's almost as if the observer is part-creating the world that is observed. While this is potentially a reasonable proposition for an idealist, it seems less so for the materialist. If matter isn't solid and fixed, then it's not really matter, is it?

This theory offers a more down-to-earth explanation of the effect of observation on reality. It doesn't "create" reality, it just quantifies it.

Observation is equivalent to quantification.

Instead of suggesting the world isn't fully formed until it's observed, it merely states that it hasn't been observed. *I.e.* it hasn't been quantified *measured* known / seen. The world always exists fully formed (relative to us) as it is created prior to our existence, but it is made of qualities and quantities (observations) are a secondary, derived phenomenon.

The world is always in a solid form even when we don't observe it, because "solid" is a quality.

Observation adds a new level / layer of detail to reality, "transforming" it from being qualitative to quantitative, one-dimensional to two-dimensional, absolute to relative. It provides a new perspective, adds new detail, and allows things to be described and communicated. It makes reality more "real" only in this sense.

The Spirit / Matter Duality

Reality appears to consist of two fundamental complementary categories of stuff: "matter" and "spirit". The word "spirit" has a lot of mystical and religious connotations, none of which are to be implied here. The purpose here is to demystify the idea as much as possible and make it a term we could use in a scientific context.

I suggest the following definition as a start:

spirit
- anything we perceive as a non-physical object: e.g. universals, abstract concepts

matter
- anything we perceive as tangible and solid, not spirit

When we talk about the "spirit of the law" for example, we're discussing its purpose / intent. Intent is something intangible, it's made of spirit but it's not at all mystical or beyond human understanding. Desire is an everyday, ordinary phenomenon that we all experience and are familiar with.

Everyone understands the concepts of "will" and "laws", there's no mystery there. Physical matter, on the other hand, is extremely mysterious. What proportion of people understand Einstein's theory of relativity, for example, or the nuclear "weak force"?

The spirit-world is far easier to understand than the world of matter. Yang is simple, Yin is complex.

It's ironic that spiritual phenomena are considered "mystical" and "unscientific", when they are the things we experience directly and can personally know, whereas the claims of material science usually rely on third-party opinions and beliefs. (People will believe experts, but not their own senses.) It's also ironic that some imagine the spiritual to be something beyond logic or comprehension, when it is in fact "knowledge".

The spiritual is not mystical. We're all familiar with desires, we all have them. Desires are intangible, but they exist, and we can perceive them. Other examples of "spirits" are information, rules, emotions, thoughts, ideas, wishes, pain, and wisdom.

Abstract concepts like "dog" are spirits. The dictionary is largely an alphabetised list of spirits. The "spirit world", at least in this context, holds no mysteries, it's an open book, none of it is hidden from our sight.

The counterpart to spirit is matter. These things are a duality and must be considered as being relative / in relation to each other. Spirit is not-matter. Matter is not-spirit.

A living organism is "made of spirit and matter". It has a body, but the body is driven by the intent of the creature. It's the will to eat, mate, build shelters and so on that drives activity. These intangible desires may be entirely generated by the body's material needs; a "spirit" doesn't have to be at all "spiritual". Hunger is a spirit, a motivation, as is disgust or lust. These things move you into motion, they are intangible motivators.

Materialists will argue that such "spirits" could be created by physical processes, but this theory says we should categorise phenomena as they are perceived. We should "trust our senses", by which I mean we should analyse all systems based on the perceived qualities they have.

The "spirit-world" by this definition is the conceptual-framework, the "world of ideas", and we are generally familiar with its content. It's not mystical or beyond human comprehension. The spirit-world is fully open, simple, and transparent. It can be known directly, relatively easily. It's the opposite of the material world, which is complex, hard to understand, and can only be known indirectly.

Another definition

While the above definition works in some contexts there is a better one. Spirit is Yang, and matter is Yin. That is really the ultimate, perfect definition.

Spirit = Yang
Matter = Yin

In other words, before we can understand matter, we must understand duality and the nature of Yin.

"Matter", in the most general sense, really should include all the Yin properties. Anything that's "solid" in any sense including "importance" can be thought of as matter. Problems "matter", they are a form of it. When we think about a topic, there is a body of "subject matter" to consider.

The fact that these two meanings are combined in the English language is most convenient.

matter
- a subject under consideration
- the substance of which a physical object is composed

Archetypes and Symbols

As a different way of viewing the spirit / matter duality, we could reframe it as the duality of archetypes and symbols.

archetype - an original model or type after which other similar things are patterned; a prototype

symbol - something that represents something else by association, resemblance, or convention, especially a material object used to represent something invisible: "the lamb is a symbol of innocence."

The archetype is the original pattern or idea of the thing, and the symbol is a representation of the archetype, an instance of it. The archetype is the spirit, and the symbol is the matter. The concept of "hot" is an archetype, things made of matter which are hot symbolise it.

An archetype is a template, blueprint, prototype, or plan. A symbol is a representation, product, example or instance.

Plato's Forms

While "Platonism" is taken to mean any view which regards abstract objects as real, his description of them unfortunately seems to be a paradox. Plato's conceptualisation differs from this theory in that his "forms" were described as the "perfect example" of a thing, where, "perfect" is taken to mean "flawless", or "without error".

Perfect

A "perfect example" in this sense would have to be very specific, but in the UP, archetypes are very generalised categories, which is the opposite view. Plato's forms are specific, the UP's are general.

While Plato's idea of the archetype of "dog" would be an idealised "perfect dog", in this theory it's all kinds of dogs lumped-in together, it's a category or a "super-concept". Adding the word "perfect" (in this sense) to a description would make it more specific, so less fundamental.

The idea of a "perfect dog" is paradoxical. Either it would have to be perfect in form or in function, but there is no single form or function for dogs. They come in many shapes and sizes and perform multiple functions.

Complete

The word "perfect" can also mean "complete" or "whole", and archetypes are perfect in this sense. The archetype of "dog" is "complete" in the sense that it contains all dogs. In other words, the archetype is lacking nothing in descriptive ability.

To take it as meaning something like "the best dog you can imagine" would be incorrect.

Example

There's a worse problem with the word "example". To say an archetype is any kind of example is to confuse the container with the contents. The idea of an archetype being an example is a paradox. An example is a symbol, which is the opposite of an archetype, so an archetype cannot be an example. Only something instantiated in matter can be a "perfect example" of a principle.

Archetypes do contain a description of a "perfect example" of their principle, but they must also include a description of the "worst example" as well. The archetype must be able to contain all possible instances of its contents. An archetype is more like a category defined by a list of properties.

The Ghost in The Machine

At all levels of the conceptual framework there are ideas about non-physical things. A measurement is driven by a desire, and its purpose is to find out information. These things are immaterial and intangible. Matter can convey information, like a newspaper can convey news, but the paper and ink aren't the information itself. A consciousness must interpret the symbols on the page for them to have meaning.

Without the observer, all information would be meaningless. The observer is the "ghost in the machine", a "spirit" and it must be there to interpret and understand the information. According to the logic of our conceptual framework, there must be a consciousness of some type to take a measurement / experience reality. There must be both the thing to be measured, and a measurer to do the measuring.

The measurer must want to take that measurement, for some reason. There must be intentionality in the equation, a reason "why". Machines can take measurements, but they don't know why they're doing it. Machines must be made by engineers who do know why.

While some philosophers may argue that consciousness and intentionality is an illusion created by the brain, our conceptual model requires the idea of intent for it to make sense. A measurement being taken without a reason is impossible and unimaginable.

We can imagine an experience being had without intent as experience is a passive concept, but it still requires an intangible experiencer. There's a "ghost in the machine" and we can't easily exorcise it, because it's us.

The Unitary and Multiple Sets

The concept of duality has more internal structure than is commonly recognised. It contains its own facts, rules, and a long list of fundamental properties.

Duality contains two distinct categories containing related concepts: the singular set, and the plural set, one and many, Yang and Yin. The concepts in the two subsets are all linked and have some mutual relevance. They help to define, to provide context and meaning for each other.

- Concepts in Set (A) correspond to the observed, and the number two / multiplicity.
- Concepts in Set (B) correspond to the observer, number one / unity.

While some concepts in each set may not seem to go together at first, with a broader interpretation it's usually possible to find a significant link.

Concept / Property	Set (A) - MANY / Matter (Yin)	Set (B) - ONE / Spirit (Yang)
Knowledge	"Known", *i.e.* Observed, Measured	"Unknown", Unmeasured, Observer
Substance	Matter, Tangible, Noun	Spirit, Intangible, Verb
Data Type	Quantity, Discrete, Digital, Countable	Quality, Continuous, Analog, Measurable

Archetypes	Symbol, Representation	Archetype, Template
A Construction	Components	Connections
Life / Consciousness	Dead, Unconscious	Alive, Conscious
Role	The Observed	The Observer
Time	The Past	The Present, Future
Distance	Far, Distant	Near, Close
Activity	Passive, Object	Active, Subject
Relationship	Relative	Absolute
Precedence	Follower, Second	Leader, First
Strategy / Approach	Opposing, Competing	Complementary, Cooperating

This theory proposes that these two sets are distinct and real "proto-concepts", or "super-concepts", and they are the foundation from which all other concepts are derived.

proto-
- first, beginning, giving rise to
super-
- over, above, beyond

They are like super-sets which contain all other concepts, but more than that, they are concepts in their own right. We can link all the ideas in each set together:

- Matter is known, it's tangible and is measured / observed. It's a symbolic representation of an archetype, the archetype comes first, matter second. It's "dead" and passive and appears as discrete objects. Matter is what you observe and is (relatively) "far away" from you, and so on.
- Spirit is "unknown" in the sense that it is an unmeasured quality, or is the "knower". It's alive, it's close (i.e. it is you). It is the active observer, the reference point by which everything is ultimately judged (the absolute).

We'll briefly discuss some of the connections between concepts to get a feel for how they relate.

Connections are alive?

In a construction, like a house, the components are many, and there's only one "right" way to connect them together. So those concepts go together easily, but the connections in a house-build aren't "alive", are they? Why are these ideas connected?

When a house is built, the components (bricks, tiles, timber etc.) are pre-made, in the past, but the connections between them must be made on site, in the present / future. The connections are the part that is the active focus of work. Just like a TV channel can be said to be "live", the connections in a construction are the active part that is "being done", it's in-progress and "live" in that sense.

To understand connections between concepts it's necessary to think very broadly. We may have to consider alternatives meanings of words, original etymology, or abstract the spirit of the word so we can re-phrase it.

Connections are qualities?

The list above correlates connections with qualities, whereas quantities are correlated with components. Why are connections "qualities"?

Consider a brick wall. Before it's built it's just a pile of components, it has no qualities as a wall because it isn't one yet. Only after all the bricks have been joined together can we attribute any qualities to it, like "ornate", "tall", "solid" or "well-made".

The connections turn a pile of components into a thing that has qualities, including the quality of quality, the attribute of how well it was made. Quality (good/bad) is the most basic quality a thing can have.

quality
- an inherent or distinguishing characteristic; a property
- degree or grade of excellence.

It's the quality of the connections that defines the quality of the product. All house builders have access to the same bricks and cement, but some make better quality houses by connecting the parts together better. It's the same with any form of construction, the quality of a thing is defined by how its components are connected.

Al music is made from the same notes, but the quality is infinitely variable.

"Yin and Yang"

Duality exists, and it is quite easy to perceive. It is a real phenomenon, and people of many cultures throughout history have noticed it. It's relatively easy to discern that everything that exists is made of duality, so it is no surprise that cultures have attributed divine qualities to it and its two components.

What should we call the two halves of duality?

The two sets (A) and (B) are like conceptions of duality in various philosophical or religious schools of thought from around the world, including:
- The Taoist concepts of "Yin and Yang".
- In Alchemy, "Moon and Sun", or "Silver and Gold", probably others.
- Maybe "Heaven and Earth" from various traditions (e.g. "Ouranos and Gaia").

"In the beginning God created the Heavens and the Earth"
Gen 1:1

The first line of the Bible quoted above could be referring to the origin of the concept of duality, rather than anything physical. The "heaven *earth*" pair is *potentially just another naming convention for Yang* Yin.

The Alchemical naming convention of "Sun and Moon" is quite a good one as these two celestial bodies do embody some properties of the two principles.

Moon - Yin	Sun - Yang
Reflected Light - Receiver	Light Source - Provider
Cold *Blue* Silver	Warm *Yellow* Gold
Many shapes	One shape
Night / Dark	Day / Light

I don't want to use these words however, because I feel they're ambiguous, I'm not referring to the actual sun or moon. The two parts of duality need their own dedicated names.

I'll use the names "Yin" and "Yang" to refer to the two parts of duality. They're short and unambiguous, and there is no better terminology I can think of. Unfortunately, we do not have any words in English that represent these two entities, so I'd have to make some up and I'm reluctant to do that.

I don't want to have to "reinvent the wheel" when the popular understanding of Yin *Yang is reasonably accurate. They are commonly understood as referring directly to the two elements of duality, often as the "feminine" and "masculine" principles respectively. I'll be writing them like proper names (capitalised), as if they were people* minds.

Yang is the set of unitary concepts: one, single, individual, whole, absolute; it is the "masculine" principle. A human male cannot become pregnant, he is always one and never two people. Yang is one.

Yin is the set of multiple concepts: two, many, group, separate, relative etc.; it's the "feminine" principle. A woman can carry none, one, or more children, she is sometimes two or more people. Yin is many.

The Sword in The Stone

I suggest that both these symbols were designed to represent the duality principle.

Some Europeans reject the concept of Yin / Yang because it's considered to be an Eastern idea and irrelevant to other cultures, but while the names may be Chinese, the underlying concept is universal.

The principle of duality really is the foundation of all thought. The fact that English doesn't have its own words for Yin / Yang presumably explains why English-speaking cultures don't have a strong understanding of it.

It appears however that the principle of duality may well have been common knowledge worldwide in the past, including in the west; there are clues it was an important part of our culture. It could explain why many European languages have (or had) gendered nouns, for example.

Some important European cultural symbols may benefit from being considered in this light. The "Sword in the Stone" in the legend of King Arthur could be a duality-symbol. *I.e.* the stone represents Yin *Matter, and the sword is Yang* Spirit. If interpreted in that way, the story then becomes an allegory about virtue.

In this case, King Arthur is figuratively demonstrating his virtue by being able to extract the sword (his spirit) from the stone (matter). He separates himself / his spirit from the urges of the physical body. He has attained "self-mastery" and has overcome his ego. His will is no longer controlled by his body, the reverse is true. So, I suggest these symbols are somewhat equivalent, they both refer to the archetypes of Yang and Yin.

Excalibur

The etymological dictionaries suggest the origin of "Excalibur" is:

Excalibur:
- King Arthur's sword, c. 1300, from Old French Escalibor, corruption of Caliburn ... apparently from Welsh Caledvwlch probably a variant of the legendary Irish sword name Caladbolg which might mean literally "hard-belly," i.e. "voracious."

https://www.etymonline.com/word/excalibur

It seems strange they would miss the following rather obvious breakdown:

ex-: out of
caliber: The inside of a hole (e.g. gun barrel / mold)

"Excalibur" then simply means "out of a hole", which fits the story perfectly, and has profound meaning. Perhaps that interpretation was considered just too ordinary by the editors, if so, they rather missed the point.

The underlying archetype of "out of a hole", is "spirit escaping matter", "leaving darkness for light", "freedom", and "enlightenment. It is a deep, meaningful, and poignant archetype describing the ultimate mission and destiny of humanity.

People tend to miss the significance of the simplest concepts.

The "Lady of the Lake"

There are in fact two legends of how King Arthur obtained Excalibur. There's the "sword in the stone", and also "the lady of the lake". Historians aren't exactly sure why this is the case, but if it is a duality symbol then we can interpret both stories as being essentially the same and answer that question.

The "lady of the lake" is the "lady of the lack", Yin.

Yin is the archetype of "absence", the "hole" or "void". In this case, both stories are describing the withdrawal of Yang from Yin, spirit from matter. We'll come back to this idea.

Some Common Dualities

Here's a table showing some common dualities divided into the two categories.

The Yin category is associated with the "left", Yang is "right" so they're shown in this order. In the duality right / wrong, Yang is right. In some religious / mystical systems there is the idea of a left and right-hand path, which also corresponds to this arrangement.

There is congruence between the items in each list, and they make useful connections between ideas. However, it's not always easy to categorise ideas, even into just these two categories, it can take some thought.

Yin, Negative, Many, Moon	Yang, Positive, One, Sun
Passive, Indirect, Specific	Active, Direct, Generalised
False, Imaginary, Fictional, Inaccurate	True, Real, Factual, Accurate
Incomplete, Partial	Complete, Whole
Receiver, Direction = Inwards	Provider, Direction = Outwards
Child, Dependent, Immature	Parent, Independent, Mature
Matter, Tangible, Defined, Limited	Spirit, Intangible, Undefined, Unlimited
Slow, Static, Fixed, Formed	Fast, Changing, Formless
Heavy, Shadow	Light
The Observed, Matter, Body	The Observer, Consciousness, Mind
Emotion, Feelings	Reason, Intellect

Time

The phenomenon of time provides a simple example of how relationships are relative, and how something can be Yin in one relationship and Yang in another.

If we consider just past / future, then future is Yang (unknown, active) and the past is Yin (known, static). Time is thus "driven by" the future, it's the active principle in this relationship. Everything we do is driven by our desires for the future.

If we include the present however, both past and future are Yin compared to it. Past and future are less real than, and dependent on, the present. This indicates the present exists at a level above them in the hierarchy.

This relationship is the same pattern as we saw earlier with the two observers. Here the present corresponds with One, the future with Yang, and the past with Yin. The present drives creation from "outside" the universe, the future drives it from within. We'll come back to "time" shortly.

Yin, Passive, Many	Yang, Active, One
The Past, Before	The Future, After
The Past, and The Future	The Present

Relationship

A thing can't be Yang or Yin alone, it must be in relation to something else. I found this sentence repeated on many websites:

"Wheat in the field is Yang, but once it is harvested, it becomes Yin."

This begs the question, what is the wheat in the field Yang relative to? Once harvested, what is it Yin in relation to? Without closing that loop, the statement is meaningless.

Something may be Yang in comparison with one aspect of a thing, and Yin in comparison to another aspect of the same thing.

For example, a child is Yin to their parents, as they depend on them for physical survival, but they are Yang to them in the meaning they provide. The child is what makes "family", which is what the parent desires. A child is thus Yin to their parents in the material relationship, and Yang in the spiritual (meaning, purpose).

Duality implies everything is made of reciprocal relationships like this, which are equivalent to circles, loops and "circuits". It suggests all processes are analogous to electrical circuits.

Anything that is the provider in a relationship is Yang, its receiver is Yin. Plants are Yang to animals when acting as a food source for them. They also provide the oxygen they breathe, homes to live in *etc*. But plants are Yin to the animals' manure and the CO_2 they exhale. It's the "circle of life".

Inside / Outside

If we return to the idea of the two types of dualities (opposing and complementary) and write down their properties, we find an apparent anomaly.

A complementary duality (driver/car, teacher/student) is two things coming together to form one thing. It's a "many-to-one" process, unification. Complementary dualities are a cooperation, harmony, which is also Yang. However, the number-line it generates is discrete *quantised* digital, which is Yin / Many.

Parent / child is a complementary duality and it's either true or false. You can either be a parent or not be one, there's no spectrum of possibilities. So, complementary dualities (Yang) look like discrete units (Yin), they're "on / off", but opposing ones (Yin) look like a continuous range (Yang), a spectrum. They're created by division of one into many, which is Yin, but they appear to be continuous, which is Yang. How do we reconcile this?

Simply, the two qualities of "how a thing is made" and "what it looks like" are not the same. In fact, they're opposites and we can equate them to the duality "inside / outside". We can view these ideas from the inside, *i.e.* what they're made of, and from the outside, how they appear.

We're skipping ahead a little here as this is a two-dimensional relationship, and these are defined at the next level down the hierarchy, at "Water". This seems like a good place to introduce the idea though. (The two dimensions here are "oppose *complement*", and "*inside* outside".)

What we find is that the inside and outside aspects of these concepts are the opposite, and this must be a general rule in all 2D relationships (this is considered a principle in Taoism):

What's Yin on the inside is Yang on the outside.
What's Yang on the inside is Yin on the outside.

The concept of inside / outside acts like a kind of mirror, as "reflection" is the primary mechanism in the UP. They are the inverse of each other. Of course, this isn't to be interpreted simplistically. Something that's physically large on the outside isn't physically small on the inside, but it will be small in another sense.

Small on the outside, large on the inside?

How can we interpret this? We should be careful in applying this rule, especially to 3D objects where it may not apply. However, a possible example might be as follows.

The rule says: a house that is physically small on the outside (what it looks like), is large "on the inside", (how it's made), and vice versa. We can define "small" as "more limited", and "large" as "less limited".

Essentially, a small house is "large" in the number of ways it can be built, whereas a large house is "small" and more limited that that regard. There are many ways you can build a small house, such as a cloth tent, a mud hut, a wooden cabin and so on. It's easier to build a small house and you have more flexibility in how you do it. You don't need strong structural components to span large rooms or multiple floors, so there aren't many limits on what materials you can use.

A house that is physically large on the outside is small ("limited") in "how it's made". There are fewer ways you can build a large house, as physics imposes natural limits on what is possible. You need strong structural components to build a multi-storey construction, you can't make a skyscraper out of tent material or cow dung.

Smaller buildings are thus "larger inside" because there are more ways to build them. They are also easier to maintain, and easier to change. "Easier" generally correlates to "less limited".

Student / Teacher

Consider the complementary relationship of a teacher and student. If the above rule is applied, we get the following.

1. The teacher is a student inside.
2. The student is a teacher inside.

Again, "inside" means "what it's made of". It's content.

Statement 1: a teacher is "made of" studying. To become a teacher, you must first be a student. The information you learn as a student is what you teach as a teacher.

Statement 2: if the "student is a teacher inside". We can interpret "inside" as "desire" or "in potential". A student is "made of" the desire for knowledge. The student desires to learn, and they are a potential teacher.

I'm interpreting "inside" / "made of" as the following things, is this valid?

1. Past learning.
2. Desire, what the aim is.
3. Potential, what it could be.

Why is there more than one meaning?

These are complementary relationships, so are mirror images. If we add time into the equation, we can see they obtain their properties from different directions in time.

The teacher is "made of the past": their time spent studying. The teacher was a student in the past. The student is "made of the future": their desire to know things and succeed in life. The student is a teacher in potential / the future.

Child / Parent

A parent was once a child, they are "made of childhood" (the past).

The child is a parent in potential, they are "made of potential" (the future).

Property / Owner

If you've ever had pets, you'll know that despite officially being their "owner", in some significant ways they own you. When you have pets, you must consider their needs. You can't go anywhere without making arrangements for them.

Any significant property that requires maintenance is something of a burden. To own a thing is to be responsible for it, to have a duty towards it. So, while on the outside it looks like you own the property, on the inside it owns you, or at least it owns some of your time and attention.

This exemplifies a paradoxical duality. "The more property you have, the more it owns you".

- The bigger your house is, the more effort it takes to build and maintain it.
- The more stuff you buy, the less room you have.

We'll explore this in more detail later. Generally, we need to be aware of the dimensionality of a concept. Some ideas are 1D, some are 2D, and combinations of 2D ideas create the third dimension.

Time

Time appears to us as a one-dimensional feature of reality like a number-line that goes from past to present and on to the future. The present is our current measurement on the ruler of time (the "point"), with the past behind/left and the future in front/right of us as parts A and B of the ruler.

The known/defined part of the ruler (A) is equivalent to the past, it had a beginning (birth) and an end (now), it's bounded at two ends. The future is the undefined part (B), bounded only at one end. The present is a strange and elusive thing. As we try to grasp it, it's already gone; like the measurement point it has zero size.

Quantisation

It's not possible to process zero length time segments, so the mind must quantise experience into discrete chunks or "frames" for processing, a bit like how a ruler has gradations. In theory a ruler is a continuous measuring device capable of arbitrary precision, but in practise you wouldn't use one to measure anything less than a millimetre, and you'd quantise any result to that level of detail.

quantise
- to limit the possible values of (a magnitude or quantity) to a discrete set of values

This is another profoundly important concept.

As we shall see, all information is created by this process. The essential underlying archetype of information is as a "quantised continuum". A "continuous object divided into relatively arbitrary quanta" is arguably the most fundamental description of reality.

Reality is created from a "quantised continuum".

We will come back to this.

Spacetime

Time and space are considered to be a single four-dimensional object in Einstein's theory of relativity, but the archetypes suggest this is not conceptually valid and offer an alternative view. The concept of time is quite different from space. We certainly don't enjoy the same level of freedom of movement in time as we do in space. On this basis alone they seem to be different classes of thing.

If we find we like a particular location, we can stop there and enjoy the view, but any particular moment is already lost to the past as it reaches our attention. While we can consider both space and time to be dimensions, there's an aspect of them which is opposite or complementary. One offers freedom, the other is more like a limitation, a road we are forced to walk.

Another oddity with this view is that the "present" has zero size; it's a point with no dimensions at all, and yet it is the most "real" form of time. We can view past/future as a dimension, but it's something of an illusion. The three dimensions of space exist in the present, but past and future (by definition) do not.

Space / time is a duality.

The UP offers a different view on these phenomena. It says that time is a "mirror image" of space. Space / time is a complementary duality equivalent to "distance *change*", corresponding to *Yang* Yin respectively.

Strictly we should frame the relationship as distance (Yang) vs "measurement of change in distance" (Yin).

space: distance (Yang)
time: measurement of change in distance (Yin)

It turns out that both space and time have three dimensions that exist in the present, as will be discussed later.

From Past to Future

We consider the past to be "behind" us, and the future "in front". When we walk, we usually walk forwards. Our eyes are on the front and our senses of hearing and smell focus that way too. When we walk, what's ahead of us is our future location, and what's behind is the past.

The past exists in the present in the form of matter. Matter is "rooted in the past"; history is encoded into it. We can attempt to deduce what happened in the past from the way matter is configured today, but it's difficult. We could view matter as "the past in encrypted form".

matter
- the product of past activities, a cryptographically encoded history.

Our material bodies are concerned with the past. Their current condition is determined by history. They are busy digesting food eaten in the past or recovering from injuries sustained.

The future, on the other hand, only exists in minds. Minds are "rooted in the future", that's "where they live". Providing for the future is the focus and purpose of all conscious work.

The future can be changed, or at least that is how we perceive it, but the past we have already seen, and it can't be changed. The past is fixed, it's literally "set in stone" in the universe around us. In the future though, anything could happen; you might win the lottery!

The past is made of solid facts, but the future is made of intangible hopes and dreams.

Past - Yin - Matter	Future - Yang - Spirit
Behind	In Front
Fixed, unchanging, known	Undefined, unknown
Facts, memories	Hopes and dreams
2-ended: from beginning to present	1-ended: from present to infinity
The body lives here	The mind lives here

Precedence

Physicists wonder why time flows in only one direction, as their equations allow it to run either way. The is the mystery of the "arrow of time", why does it point toward the future?

Duality answers this question succinctly and in many ways. There is always a natural, logical, sequence which things must follow.

precedence
- the fact, state, or right of coming before in time, order, or position

This is related to the concept of contingency.

contingent
- dependent on other conditions or circumstances; conditional:
synonym: dependent.

We could rephrase this as a natural "law of precedence".

law of precedence (contingency)
- a thing must exist before other things can be derived from it.

The rules of precedence which are built into the concept of "two" are the reason for the "arrow of time", why time moves from past to future. I think this eliminates the possibility of physical time travel, at least within the simulation.

The universe is an iterative process, a series of contingent events. Future states are derived from past ones. Causality occurs in discrete steps / events.

Some things must come before others.
- The cause must exist before the effect.
- The parent must exist before the child.
- There must be light to create a shadow.
- You must be alive before you can die.

Some dualities, such as the above, obviously contain a dominant / active member and a subordinate / passive one. The child didn't ask to be born to the parent. The parent is the active party, the cause. These are all cause-and-effect relationships.

Causality is the physical manifestation of the "law of precedence".

Hard / soft

Some dualities don't seem to have any particular order. Hard and soft for example. Which of these concepts comes first? There's no obvious precedence in the concept of "hardness", but then it's not a property with any sort of time component.

However, because we can correlate these ideas to Yin / Yang, we can assign a precedence to non-temporal concepts. All dualities contain this relationship, even if they don't have a time-component. It works out that "hard" comes before "soft". A receiver must be soft, so it can receive. A provider is "hard".

Spirit is hard, it can't be changed and it's permanent (consider the "laws of logic"). Matter is soft, it's always changing and is temporary. Spirit is made of an indivisible, single thing. Matter is made of many things which can be divided.

This corresponds to the idea of a "big bang" which, if true, would have been the "hardest" thing (explosion) ever, and explosions are "made of energy" which is "spirit". An explosion like a bomb would be an unlikely interpretation of the archetypes though. The UP suggests the universe grew like a tree.

Time Is Change

The only way to track the passage of time is by tracking changes in something, the sun and planets, pendulums, clockwork, quartz crystals, *etc.* Time is a measurement of change, so time is conceptually equivalent to "change". "Time passes" means "change happens".

We should consider "distance" and "change" as two parts of a complementary duality. They are equivalent to "space" and "time", and they are the most fundamental properties of all reality, not just the physical. Time is a measurement of "changes in space", so "space" must come first, and "time" must follow. The former is therefore Yang, the latter Yin.

I am led to believe that the SI units used in physics can all be reformulated in units of space and time (meters and seconds). This would mean that the units for mass, voltage *etc.* are not fundamental. More on this later.

The "Beginning of Time"

A popular view in physics some years ago was that "time began with the big bang"; however, the general concept of "time" is an absolute and cannot have a "beginning". It's a paradox, a category error.

category error
- a semantic or ontological error in which things belonging to a particular category are presented as if they belong to a different category

Skipping ahead a little for clarity: absolute, unmeasured time originates in the "One" category, whereas relative, measured time originates in "Heart". These are the specific categories which are being confused in this case.

If we're building a framework of ideas, the concept of "time" would have to be created before the idea of "beginning" can be. A "beginning" is a specific type of time, it's an "event". Beginnings exist within time, there must time passing before something can begin.

Conceptually, "time" is the same thing as "change". A "beginning of time" is like saying "when things began to change", implying there was stillness / no-change before the beginning, which is paradoxical.

So, time cannot conceptually have a beginning, it's a paradox.

What we find is there must be two conceptually different forms of time: "real time" and "measured time". Measured time can have beginnings, real time cannot. We can talk about the "beginning of the day" with no problems whatsoever, but we can't have a "beginning of time" in general.

Real / Measured Time

The phenomenon of time "comes in two forms": real and measured time, corresponding to the One and Heart categories.

You might think these things are the same or the difference is unimportant, but the relationship between these two views is profound. They are opposites, but without an understanding of duality we'd probably never have noticed.

Measured Time (Heart) Yin - Circular	Real Time (One) Yang - Straight
Passive - Dependent	Active - Independent
Relative, Arbitrary, Fictional	Absolute, Real
Quantified, Limited, Defined, Measured, Known	Qualified, Unlimited, Undefined, Unmeasured, Unknown
Finite, Mortal	Infinite, Eternal
Quantitative - Two-part / dimensions (i.e. number + units): *e.g.* "7 days"	Qualitative - One-part / dimension "Ages" or "Soon"

"Yang Time": Real Time

"Yang-time" is real / absolute *qualitative* direct time. It's "time passing". It must have always existed because it cannot have had a beginning. All things (including One) experience it directly. Yang-time is the straight "arrow of time" going from past to future, from cause to effect.

Yang-time is real but it's unquantified and is essentially unquantifiable because there's nothing to compare it to. It has no beginnings or endings, it's just "change happening". In this view, before the physical universe began, change could happen (time passed) but there was no yardstick, no "year" or "day", no units. There was no way of telling how much time had passed.

Yang time is unmeasurable because any division of a continuum into discrete units is necessarily arbitrary.

This is perhaps why "God" is said to be "outside time". It's not that time doesn't pass for him, it's just that there is no way of measuring how much time has passed from that frame of reference.

Far from being "timeless", it works out that "One" essentially is time. One is correlated with activity and change, and change is time. "God" / Yang is change (active), and Yin is the measurement of that change (passive).

There's only one real phenomenon of time passing, but there are infinitely many ways to measure / quantify it.

"Yin Time": Measured Time

"Yin-time" is cycles *the (sine)wave, "cyclic time", or "fictional-time". It's indirect and circular* self-referencing, as all measurements are. It has a beginning, an end, and is quantified, but it's relatively arbitrary. This is the form of time that would become possible at the "big bang", *i.e.* when things began to be separated and could be compared with each other.

(Note: the official etymology of the word "sine" does not go this route, but "sin" is correlated with the moon via the Mesopotamian goddess "Sin", and with the left-hand and darkness (sinister). The phases of the moon may have been the original inspiration for the name "sinewave". "Sin" also sounds rather close to "Yin", so I wonder...)

The length of a day on Earth is anchored in the perception of the sun's movement across the sky. It is a "relatively real" cyclic-time unit, but the day-length on Earth is arbitrary from the perspective of the universe. A 24 hour "day" is "real" on earth because it corresponds to its physical rotation, but not so real anywhere else. It's Yin-time, relatively arbitrary and "fictional", but it does still allow measurement that is useful in the local frame of reference.

Cyclic time is essentially just a "convention"; It's something people have agreed on, and/or nature has chosen, apparently arbitrarily. The measurement units we use have no significant universal meaning but can have significant local meaning.

Our senses can only work with Yin-time.

It's important to note that all our senses are dependent on Yin-time, they are measurements. The colours we see are due to measurement of the frequencies of light by the eye. The sounds we hear are measurements of sound waves by the ear. While time really flows in a straight line like an arrow, it's not possible to measure that, we can only measure "second order" time, and that appears as cyclic phenomena.

Real / illusion

The relationship between these two perspectives on time is profound and revealing. It shows the "characters" of the two parts of duality very well. Yang-time is the actual passage of time, and Yin-time is "just" a relatively arbitrary measurement of it. The former is real, the latter is more like an illusion, but we can only "know" time via this "illusory" route.

This is the deep underlying nature of the universe. Yin is a necessary (but resolvable) paradox.

Both perspectives / concepts are essential components of reality. We need time to pass, but we also need to be able to measure it, and these two concepts are necessary complements. The relationship between them exemplifies the sheer elegance of duality. it is a beautifully simple, yet incredibly powerful concept.

Hopefully this clarifies our definition and understanding of the phenomenon of time and demonstrates the utility of thinking in terms of Yin/Yang to improve organisation of thoughts.

Classifying Dualities

Some dualities are easy to classify as Yin/Yang, but I've had to re-think some over the course of the investigation. Near / far is quite easy. There's only one place which is extremely "near" to you, it's your exact location. There are many places which are far-away. So, "near" is Yang, "far" is Yin.

"Near" is the same as "here" in this context, and "here" is the active principle, consciousness, you. You are at the centre of the universe, observing it.

Everything that isn't you (your consciousness) is "far away" relatively speaking. This also applies in the context of time; the present is near, and the past and future are far-away.

Large / Small

When attempting to classify this duality it's tempting to correlate large with Yang. We might argue:

- males are usually larger than females
- Yang is "adult" and "provider", and those indicate a larger size
- One is Yang, and is "everything", so must be big.

However, if a thing has an inside and an outside, we need to bear in mind that one is the opposite of the other. The fundamental property exists on the inside because it's Yang, the secondary one on the outside. We need to make sure we're not "judging the book by its cover".

If human males are larger than females on the outside ("what it looks like") then they are presumably smaller on the inside ("how it's made"). An example might be that biological sex is determined by (relative) gamete (sex cell) size, and small gametes are associated with the male. There are other ways in which males may be "smaller on the inside", but that's a topic for another day.

The number one is Yang, and two is Yin, and one is, at least numerically, smaller than two.

"Near" *"the present"* "consciousness" is a point of zero size, things don't get much smaller than that. The observer is small, the universe is large.

So, Yang *creator is small, and Yin* creation is large.

You might ask, how then can One contain the universe? Surely the container must be bigger than its contents? We'll answer this question shortly in the section titled "The Belly of the Whole".

Infinite / Finite

There seems to be some confusion about the concept of infinity even among mathematicians.

I believe that the definition we arrive at here is a significant improvement on the current one. With careful treatment of the concept, we can remove all hint of paradox from it.

The dictionary defines the concept as follows.

infinite
- literally "without end", unbounded, limitless
- impossible to measure or count, incalculable

There's an obvious duality of finite / infinite, so how do we classify it?

- Infinity can contain a finite quantity, but not the reverse. So, infinity must come first, it is therefore Yang.
- Yang-time had no beginning and is unmeasurable, so it is effectively "infinite".

So, Yang correlates with infinite, Yin with finite, but this isn't enough definition and we hit problems. Consider the following cases.
- The (Yang) number one is not infinite, but (Yin) "many" could be.
- Yang is "small" but infinity is "large".

These conflicts suggest there must be a 2D relationship to discover, the idea needs extra parameters to be defined fully.

Quality / quantity

To provide a clear definition of "infinity", it's necessary to include a second duality in the description. The first duality we need is "infinite / finite", the second one is "quality / quantity". People tend to think of "infinity" as a purely quantitative concept, but it's not.

Things can be infinite / finite in "number", or in "quality".

The most fundamental qualities are "space and time", *i.e.* size and lifespan.
- Physical things are finite in quality.
- Spiritual things are infinite in quality (not bound by it).
- Physical things are bounded *limited* described by qualities, like colour, shape, texture *etc.*
- Spiritual things are not bounded / described by qualities because they are those qualities.

Spirits aren't defined by qualities; they are the qualities things are defined by. So, spirit is "infinite in quality".

Countable / uncountable

Mathematics recognises two types of infinity: "countable" and "uncountable", which must fit into the description. This obviously correlates to the duality discrete / continuous. Discrete items can be counted, continua cannot.

A full definition

We end up with the following duality, with Yin mirroring both "infinite" and "quality". Yin is "apparently infinite", equivalent to a "pseudo-infinity", an "illusion" / paradox. Yin cannot truly mirror the quality of "infinite", it can only provide an illusory version of it.

Yang is "infinite in quality", Yin is "pseudo-infinite in quantity".

The table below explores the properties of these two types of infinity and tries to define them more clearly.

Yin: Matter - Finite - Countable	Yang: Spirit - Infinite - Uncountable
Finite in quality (e.g. "time and space")	Infinite in quality
"Apparently Infinite" in quantity (i.e. finite)	Finite in quantity, (i.e. "one")
Many, discrete, integers	One, continuous, real numbers
A set of individual numbers	A range of possibilities / potentials
"Tending towards infinity". Apparent / illusory infinity.	True infinity
"Countable infinities"	Uncountable. Infinitely sub-dividable. Continuum.
(Matter is) Physically bounded. Limited in, and dependent on space and time. Finite in "space and time", or more generally "quality"	Physically unbounded. Unlimited in space and time. Existing prior to it. Infinite in "space and time", or "quality"
Has size and mass which is finite, but can tend towards infinity	Cannot be compared *sized*. Infinite unbounded in that sense.

The underlying archetypes tell us:

Yang is a single continuous infinity.

Yin is many discrete items. It tends towards infinity, but never reaches it.

The opposite of "one" is "many", which tends towards infinity, but never gets there. "Many" implies discrete objects, but you can never actually have an infinite number of discrete objects. It's conceptually impossible.

"One" on the other hand is "everything", a continuum, a range. It isn't composed of discrete units, it has no parts or components, but it can be infinitely subdivided. The common conception is that the set of real numbers contains an infinite number of entries, whereas in this view it only contains one.

The "set of real numbers" only contains a single object.

The UP classifies "countable" infinities as Yin, defining them as "illusory / paradoxical". It says the only "true" form of infinity is a continuum. This classification also highlights the following relationship.

Only qualities can be infinite, quantities are always finite.

It is quite common for people to discuss countable infinities as if they were true infinities, but the UP tells us this is not conceptually valid. In essence:

There is no such thing as an "infinite number".

The phrase "an infinite number" is commonly used, but it's a contradiction in terms.

The Infinity Archetype

The properties of Yin and Yang lead us to a completely solid and consistent understanding of the principle of infinity, removing all paradoxes, and making the idea much more rigorous and useful. It reveals to us the underlying form of the concept, and it is very simple yet incredibly profound.

The basic archetype for infinity is:

A continuum is a "true infinity" (Yang), which can be subdivided into a "countable infinity" of discrete parts (Yin).

In other words:

- All countable infinities are derived (divided) from continua.
- Continua are the "host" of countable infinities.

This classification seems to offer a very clear distinction between all the related concepts, and I believe offers better definition for the idea of "infinity" in general. However it is a complex idea, and this may need revisiting.

We will expand on this model later, as this is also the basic template for "information". Everything is made of continua, arbitrarily divided into discrete parts, just like the two forms of time.

Paradox

The universe contains many apparent paradoxes, the UP can explain and resolve them all.

paradox - a statement or situation which contains two opposing / contradictory elements

A paradox is a self-contradiction, an impossible situation. The opposite of a paradox is therefore a "logically consistent statement", but not necessarily a "truth". There doesn't seem to be a snappy name for the concept, so I'll frame the duality as paradox / non-paradox.

A paradox contains opposing views, it's divided in two, and it is false. These are Yin characteristics. If we put them all together, we get this list.

Yin: Matter, Indirect, Paradox	Yang: Spirit, Dirfect, Non-Paradox
Many: large, long, curved, circuitous Many opposing views	One: small, short, straight One unified view
Paradox, division, confusion, disunity Falsehood, illusion, reflection	Non-paradox, clarity, unity Truth, reality, original.
There are many falsehoods, indirect paths	There is only one truth, direct path
Countable infinities	Uncountable infinities
Impossible, invalid	Possible, valid

These correlations are saying many things including that "Yin is a paradox", and that "Yin is impossible. We'll come back to the concept of paradox because it's an essential quality of Yin which means we can (in theory) resolve all apparent paradoxes as category errors.

This table of relationships tells us that the idea of "countable infinity" is indeed a paradox, whereas an uncountable continuum is not. It's impossible to imagine a "countable infinity" because it is conceptually invalid, like the "beginning of time".

That concludes the section on duality. A lot more could be said, but hopefully it was enough to demonstrate the reality, depth, and utility of the concept. The next step is to determine where duality could have come from.

Chapter 3

Unity

One

The purpose of this investigation is to determine how the conceptual framework is constructed. We have established that duality is a key part of the framework, but how do we get there? Where does it come from? Is duality the starting point, or is it something else?

Rather obviously, duality can't be the first concept. Two is "two ones", it depends on the existence of one. If the existence of duality (within the conceptual framework) is an objective fact, then the concept of unity must share that same status. "One" also exists.

"One" is "singularity", and modern physics says the universe began as a "singularity". The opposing views of idealism and materialism again strangely share common ground, although they may not have quite the same thing in mind when they use the term.

Given the existence of "one, the question then becomes: how do you get from that to the concept of two? What tools or processes does "one" need to have available to it to generate "two"?

Reflection

One principle we observe in nature provides an answer to this question: reflection. One can become two by reflection; and it can happen in matter like a mirror, or in a mind as an imagination or "dream".

There isn't really any other possible candidate concept for this job. "Reflection" is the only fundamental phenomenon / principle which can do what's required here. It's a key concept in our reality. It appears in many contexts from physics to psychology.

reflect
- *to throw or bend back (light or sound, for example) from a surface*
- *to give back or show a (reversed) image of; mirror*
- *to make apparent; express or manifest: "her work reflects intelligence"*
- *to think, meditate, or ponder - to reflect on one's faults*
- *from Latin reflectere, to bend back*

Negation, The NOT Operator

A reflection is "not" the original. When you look in a mirror, what you see isn't you, it's an "illusion", a "trick of the light", and an "inversion".

"Reflection" in this context is equivalent to an archetypal version of the logical NOT operation, and this is the conceptual origin of all logic and maths. This is where logic begins to be expressed in the system, because it's necessary in the process of creating two from one.

In Boolean algebra there are three basic logical operators, AND, OR and NOT. For most purposes in human-life these three operators are all we need and they emerge from the UP, as we'll see.

`https://en.wikipedia.org/wiki/Boolean_algebra`

AND and OR both take two inputs, but NOT only needs one input. So, it can do the job of creating two from one without any other data.

"Negation", also called the logical complement, is an operation ... interpreted intuitively as being true when [input] is false, and false when [input] is true

`https://en.wikipedia.org/wiki/Negation`

Binary is the simplest possible form of information, and a Boolean NOT is a simple idea: it turns a 1 into a 0 or a 0 into a 1. This isn't the only conceivable form of negation though, we can also invert qualities, like hot *cold, bright* dim *etc*. A binary computer can't invert "bright", but a human mind can.

Zero, Not

The number zero is equivalent to "none" or "not", and it can act like a mirror and as a centre for coordinate systems and maps. Zero is a form of the logic operator NOT. It's a number but it has other functions too. It's sometimes more "active" like an operator, able to "reflect things", and sometimes more "solid" like a foundation or origin point.

Consider the "all integers centred around zero" version of the number-line. The zero is like a mirror reflecting the positive integers into the negative ones. It's also a natural origin, a centre-point, and an "anchor"; the thing against which everything else is compared or is "tethered".

```
... -4, -3, -2, -1, 0, 1, 2, 3, 4, ...
```

We can call zero "naught" or "nought" which also means "nothing". Although the dictionaries don't make the connection, "not" is probably connected to the word "night". A day is made of two halves, the day and the "not-day". The Egyptian goddess of night was called "Nut", a link seems possible. Perhaps it's also related to the word "North", which is also an origin point for coordinates.

Like zero, the principle of logic is also a kind of centre and origin, but for knowledge in general. So, all these concepts are united under this "reflection" category (which turns out to be the "Voice" principle).

zero
- from Arabic sifr "cipher," translation of Sanskrit sunya-m "empty place, desert, naught"

```
https://www.etymonline.com/word/zero
```

The Purpose of The Universe

The following section, and the rest of the book, will make more sense if we can propose a purpose for the universe. It may have other goals, but this seems to be the main one:

The purpose of the universe is to explain the concept of "one".

The UP is a language, and the purpose of language is to explain things. All the concepts that exist in the framework are created as references to the originating idea of "one". It looks like their ultimate purpose is to explain what "one" is.

We tend to take concepts for granted. The idea of "one" perhaps doesn't seem all that interesting at first glance. What could "one" contain that would be worth looking at? You might wonder.

Explaining One

The idea that the universe is purposely designed to explain itself might sound a bit like wishful thinking, but it really does seem to do this. Consider what Yang and Yin are:

Yang is the set of properties that One has. Yin is the set of properties it doesn't have.

Yang is like One, Yin is NOT like One.

This very first layer of information is providing a categorisation system for one to be defined by. Yang gives us all the properties commonly associated with "God" in various traditions, such as "spirit, eternal, simple, good, masculine" *etc.* Yin has all the properties associated with "creation", and "the Goddess" in other traditions. (Duality / creation as a whole can be viewed as One's feminine partner, the "Great Yin".)

All other concepts in the framework are built on this foundation of duality and so all they can do is elaborate on it. The successive levels can only explain the basic ideas of One, Yang, and Yin in more detail.

This begs the question: why does "one" need explaining?

An obvious answer would have to be that it is a self-conscious entity / a mind, and it "desires to know itself", or it desires for us to know it, or both. Perhaps those things are the same. It would make sense of all the measuring that's going on.

The idea that "One desires to know himself" is linguistically equivalent to "I desire to know myself", it means the same thing. That might imply that God knows himself by us knowing ourselves. "Know thyself" is the sage's cliché.

A sentient "One" obviously correlates to some religious concepts of "God". It is a theme in some religions that "God wants to be known" or wants to "know himself". That does seem to be a reasonable explanation, given what we've found so far. (Note that this view is incompatible with some definitions of "omniscience".)

As "One" may (necessarily) have the properties of a sentient individual, I will capitalise the word like a proper name in that context.

The Mirror

A physical mirror is a symbol of the reflection archetype. Reflection / inversion is a logical, deterministic process. A mirror is a very simple device that's easily understood.

The archetypal process of "one" being reflected into "two" is symbolised by you looking in a mirror or thinking about yourself in your mind. Note that you cannot see yourself unless you look in a mirror. You look in a mirror specifically to "see yourself". "One cannot see himself" without a "mirror" of some sort.

Note the symbolism: you are unable to see your face directly, you can only see it indirectly, via reflection, and the image you see is inverted / reversed. One can only see himself indirectly, and "creation" is that inverted reflection; it's a logical construct depicting what is essentially "the opposite of One". The universe is NOT One, it's many.

The Real and The Reflection

With a mirror, there appears to be two of you, although only one is real and the other is just an inverted reflection.

All dualities are constructed from a real part (Yang), and a reflected, less-real part (Yin).

If reflection is the tool used to create duality, then all conceptual dualities must be constructed in this way, with one real part, and a reflection. The real part is active and comes first, the reflection is passive and follows. There can only be one original, but there can be many reflections.

Consider the duality absolute / relative.
- The absolute is like the origin of an X/Y graph, there's only one. (Yang)
- The many (relative) data points on the graph refer to, *i.e.* "reflect", the origin. (Yin)

Reference is reflection.

Note that something that refers to something else can be thought of as "reflecting" it. All knowledge is a system of references, which can be usefully depicted as "reflections". Our thoughts are references to things we perceive, they are "reflections" on them.

So, we can begin to see how an entire body of knowledge, and an entire universe, can be built up from "reflections". Reality is something like a "hall of mirrors".

The Footballer Analogy

The properties of Yang tell us that One is a spirit with no body, no "outside". Lacking any physicality, One did not have a physical mirror to gaze into. The only form of reflection available to a disembodied mind is "self-reflection", imagination, contemplation, thoughts, and "dreams".

Consider a footballer (soccer player) thinking about how he could have played better in his last game; in other words, reflecting on his performance. In his reflections he will try to imagine how he could have responded to events more productively, *e.g.* how he could have scored that goal he missed.

In his mind's eye he will see himself and the situations he found himself in during the game. His reflection includes himself and his environment. It's a two-part reflection containing an imagined version of himself, and the remembered (or projected) environment.

This is important, because it clearly explains the relationship between the concepts of One and Yang, right at the top of the hierarchy. This is a key relationship we need to clarify.

One is the footballer, Yang is his imagined self, Yin is the imagined environment.

"One" vs "Yang"

What is the exact relationship between "One" and "Yang", and also "Yin"?

The only mechanism available here is reflection, so these relationships are all created by it.

Duality, Yang and Yin, is a reflection of One. Yin is a reflection of Yang.

Because Yang comes first, it must unfold as follows: One reflects into Yang, then Yang reflects into Yin.

The footballer imagines himself, then his imagined self imagines the world he is in. It's a two-step process. The "world" part of the dream is not a direct reflection of One, but an indirect one, coming via Yang. This fits the archetypes as the "world" is Yin *indirect* following. (Note, this is reminiscent of the Biblical story of "Eve" being created from "Adam".)

In the (real) footballer's imagination, he must construct the "world" from the perspective of the imagined self, who is (re)playing the game. He must imagine himself as the footballer before he can imagine the football pitch because the environment exists relative to him. The imagined footballer (Yang) is the absolute, the origin-point, and the imagined world is Yin, the many relative points on the graph.

It is Yang, within the dream, that understands the context of the game and has the correct perspective to determine what the world should be like. One "delegates" creation to Yang.

However, even though the creation of Yang and Yin is a two-step process, the dream as a whole (duality) is a perfect (complete) reflection of One. Its two parts are "the dream" and are a logical NOT of One's unity. This doesn't quite fit the archetype of a physical mirror because the reflected you doesn't create its surroundings, but it does fit the archetype of a dream or imagination.

One is the original entity. Yang is the imagined footballer in his mind. Yin is the imagined world he is in.

One is the real self. Yang is the dream-self. Yin is the dream-world.

Yang has the characteristics of One but is not the same thing. Just as the footballer and the imagined version of himself are not the same entity. Just as the measurement point when you use a ruler is "made of your consciousness", but it isn't strictly "you".

This begins to answer the question about who we are in relationship with "God" / One but it's only a beginning. We humans do not have the consciousness or perspective of "Yang", we are further down the hierarchy. (I.e. at the bottom.)

The "non-dualist" philosophy states that all consciousness is a single entity, that "we are all God", which is true in a sense. However, the relationship between One and all other things has to be something like the relationship between you and your dreams or imaginings, or you and the reflection of yourself you see in a mirror.

When we have dreams, the character we are in the dream can be totally different from who we are in real life. Is the dream-character you? In a way it is, and in another way it isn't. It's a mixture of Yang and Yin, truth and fiction.

You are "playing the part" of the dream character, but it's not really you. If you wake up, the dream-character ceases to exist, and will probably be forgotten. You "powered" or "animated" the dream-character, it had your "spirit", but the thing that was imagined is not the imagination that hosted it.

The dream-character is just a shallow reflection of you. It's a limited representation with no knowledge of the real you, and that's the whole point of dreams, to be someone / somewhere else. A reflection of you in a mirror is just a "shallow and empty" image of you, it has no spirit of its own, no ability to act independently. It looks superficially like you but has none of the content. The process is a one-way transformation, it's non-reversible.

- Could you create a reflection of yourself (in a mirror)? Yes, easily.
- Could anyone create a living human from a reflection? No.

Reflection of this sort is a one-way street. There's no way for an imaginary footballer to become a physical one. No way for a reflected image to become a real person. At least, that is the way it looks from this perspective.

Note: a logical NOT is reversible, but some kinds of reflection aren't. There must be at least two types of negation.

Defining Logic

The archetype of reflection *negation is where logic begins to be expressed in the system. Logic must originate in One, but here it begins to take form. A reflection is a logical inversion* reversal of something, it's essentially equivalent to a logical "NOT" operator, but completely generalised.

A Boolean "NOT" is a very specific version of "not", able to handle only binary 0/1. The archetype we're discussing here is instead the most generalised version of a "NOT" conceivable, and it can operate on both qualities and quantities.

Some concepts are easy to define but logic has not been one of them, so far. Logic has always been a notoriously difficult concept for philosophers to find a satisfactory definition for. Many dictionaries simply use the synonym "reason". E.g:

- *a particular way of thinking, especially one that is reasonable and based on good judgment*

`https://dictionary.cambridge.org/dictionary/english/logic`

- *logic is the study of correct reasoning.*

`https://en.wikipedia.org/wiki/Logic`

Defining a word by a synonym is a tacit admission of failure. It's such a crucial word, such a key principle of reality, how can we not have a good definition of it?

UP theory can provide much better definitions for concepts, and this is good evidence in favour of its validity. We can define logic much more easily using it, because it provides more context and a wider perspective.

The UP defines logic as:

logic

- the rules of comparison / relationship

Logic is the system of rules that allows us to bring things into comparison, and all comparisons are relationships. Its rules allow us to compare different things and define their relationship. That is its purpose. Logic is the mechanism that created duality, and duality is relationship, which is the same thing as comparison.

I hope you agree that this is a much better definition for this crucial concept.

The Belly of the Whole

The universe exists inside "the belly of the whale", I mean "whole". There isn't any other possible location for it.

"One" does not have a dual nature, so it doesn't have an "outside"; it can only have an "inside". Only things born of duality can have two aspects. One is "the whole" and "everything". Nothing can be "outside" or separate from it as that would mean it wasn't "everything" anymore.

(Note: we can imagine "inside" without "outside", but not the reverse. So, inside is Yang.)

The only place the universe can exist is within unity, inside the "universal set". Ultimately, for the idealist, reality can only exist inside the "mind of God", or the "universal mind".

The concept of "One" must contain all the other concepts, but is this a paradox? The universe is "many". How can one be "unified" if it contains many things? This isn't a problem; a container is still a single unit, whatever it contains. Its contents don't change its nature.

Large / small

However, we discussed the duality large / small earlier on, with Yang being small, and Yin large. This means One / creator is small and creation is large. But if creation exists inside One, this raises an apparent conflict.

How can One be small relative to the universe if the universe exists inside One?

Surely a container must be bigger than its contents?

We should note that the physical property "size" cannot apply to One, because it's non-physical. So, we must use analogous concepts like "simple / complex", or "single / multiple". We can only really phrase the question as "how can one contain many?" or "how can a simple thing contain a complex thing?"

This is answered by the new definition of "infinity" we have already discovered.

Let us consider "one" to be the space on the number line between zero and one. It's a single object, an infinite undivided continuum spanning a "distance".

We can notionally subdivide this continuum infinitely, producing a "countable infinity" of discrete objects.

A continuum is simple. Dividing a continuum is complex.

This is how "one" can be small / simple yet contain large *complex* many.

Real / illusion

We can only subdivide a continuum in our imagination, as a "convenient fiction", because it has no natural divisions. Any subdivision is necessarily arbitrary / imaginary. This is Yang and Yin, of course. One is Yang *real* absolute *infinite* continuum, the subdivisions into discrete parts are Yin *illusion* relative. So, "one" can indeed host "many", but they are necessarily Yin *fictional* arbitrary *etc.*

In the footballer analogy, we saw above how the environment (Yin) is created by Yang, by effectively "projecting" it outwards. (Yang's direction is "outwards".) So, reality is created from the inside-out, from consciousness to matter, from the small to the large, where "large" is something of an illusion.

A Separate God

The Christian view is that God is separate from creation:

God exists wholly apart from His creation. He does not need the universe, and the universe is not part of His being or nature. God created all things outside Himself rather than merely emanating them out from His own essence. God is not the universe, and the universe is not part of God.

`https://carm.org/about-god/is-god-distinct-from-creation/`

In this perspective, the universe was created "outside God", but this is a paradox. If creation exists outside the creator, then it would mean that God wasn't "everything". It would effectively put creation on an equal footing with the creator, on the same level of the hierarchy.

The UP states that One is a disembodied mind with no physical component and no "outside". "God is a spirit", a mind without a body. The only place creation would be possible is within that mind. The universe must be a "dream" which exists "within the mind of God". It's an "imagining", a "thought experiment".

The Christian view suggests the universe doesn't have much relationship to God, not sharing "His nature", so it doesn't really have much to say about God. It suggests we can't know God by knowing the universe because they have different natures. This puts "God", and truth *science* knowledge, somewhat out of reach.

While it's true that "one" and "two" have different natures, they are still closely related, and we can understand the relationship between them very well. The relationship between "God" and "creation" can be known because it is identical to the relationship between the numbers one and two. The problem is these most basic concepts just aren't understood properly, their significance is completely overlooked.

The UP describes an "Integrated God" that is intricately entwined within reality and not at all separate from it. Creation is "about God", and it's "made of God" just as the number line is "made of one(s)". That is the only raw material available to make anything from.

So, The UP takes entirely the opposite view to the Christian one, *i.e.* that God can be known via creation because it is entirely made of God and is designed specifically to explain God.

The Shadow Analogy

Physical nature is a symbolic representation of archetypes; it contains many thought-provoking analogies and elegant ways of presenting the same idea. Nature tries to explain things to us in multiple formats, to help us understand. We're going to dive into some more deep symbolism here.

Imagine standing with your back to the sun, looking at your shadow on the ground.

This is yet another representation of the relationship between One, Yang, and Yin. One is the sunlight, you are Yang, and Yin is your shadow. One is the measurer, you are the "measuring", Yin is the measured / known.

Consider the symbolism inherent in it. A shadow is an absence of light, which is created by light. It takes the form of the object that caused it, but with a loss of information, converting it from 3D to 2D, and only retaining an "outline".

Your shadow is a "you-shaped hole in the light".

Your shadow is shaped like you and projected in front of you, so you can see it. By looking at the shape of your shadow and turning to the left and right, you can get some idea of what you look like. In your mind, you can "fill in" the shadow with your inferred appearance. This is somewhat like looking in a mirror.

A shadow can "tell you who you are" by showing you the "shape of the light that is missing". A shadow is like a missing jigsaw piece. It tells you the shape of the part you need to find.

This symbolises the relationship between the universe and its originator. The universe is a "God-shaped hole in the light". By looking at the shape of the hole, seeing what's absent from the world, One can potentially "know himself". He can see the shape of what "matters" to him.

This idea can be interpreted in many useful ways, not just from a "religious" perspective. To find out "what is the matter" we "shine a light on the problem".

The purpose of a hole is to be filled.

The concept of "filling holes" is equivalent to desire. All desires are "holes" we are seeking to fill.

Going back to the footballer. In his imaginings he is looking for the reason why he missed that goal. He is trying to figure out how to avoid making the same mistake again. He is trying to determine the "shape of the missing light", he's looking for the information / skills he lacks. In this case a "hole in the light" is equivalent to a "defect in my understanding", an "absence of data".

The footballer created an imaginary world to help him solve a problem, He projected his shadow / absence of skills onto the "ground of his mind", into a virtual world, where he could play out fictitious scenarios to try to find the abilities he lacked, to fill in the hole in his understanding.

We are all seeking to fill holes of varying sorts, to resolve absences, be it the hunger in your stomach, the hole in your bank balance, the empty bed, or the empty day. The universe, and our lives, are determined as much by the things that do not exist, as those that do.

In physics "energy is God", it takes that position in the UP hierarchy, and physical things in this universe have all kinds of different "energy-shaped holes" they are seeking to fill. This is why matter has structure. All forms of matter are "imperfect", they lack something, and that is why they join to make atoms, molecules and so on.

Questions to answer

The shadow represents mystery, darkness, and concealment. It symbolises things which are hidden from view, which we have questions about.

If the universe exists to explain "One", that means there are questions to answer, there is a shadow which needs to be filled with light. We will see later how question words in language correspond to the "desire" category. The conceptual framework says the purpose of reality is to answer questions. Questions are the driving-force of language, and of everything.

The best way to get a question answered is to ask as many people as possible as simply and clearly as possible, and in multiple forms. This is what the universe appears to be designed for. The repetition of an archetype in a variety of symbols is the same as asking a single question in multiple ways.

The symbolism of the shadow is quite poetic, but more importantly it's a claim about the structure of reality in general and makes predictions relevant to physics.

Holes in the Aether

Another interpretation of the symbol of the shadow, is Yin is a "hole shaped by energy", where "shaped" can mean "created" or "defined". This is, I believe, an important perspective; I think it's telling us how (physical) matter is constructed "under the hood". *I.e.* matter is "made of holes".

Specifically, it says that matter is made of holes in the "aether", which must be low pressure regions, created by energy. We'll see how this translates to physics later, however, this is a universal archetype, and applies equally to non-physical matter.

Information is the spirit equivalent of matter, and it too is made of "holes in the aether", where "aether" simply means "container" or "substrate".

Defining "Information"

This leads us to a deep understanding of the fundamental nature of information, and a new definition of it. There is an underlying, archetypal form for information.

Information is "holes" in a continuum.

All information can be represented as "items on a line", a one-dimensional string of "values".

Any type of information, with any number of dimensions, can be "serialised" into a 1D string. Consider the fact that all computer storage is effectively just a long list of zeros and ones, and it can store any conceivable type of information.

We can view digital storage as a continuum of "ones" dotted with zeros / holes.

(Note, we can never have an infinite amount of data because it is a Yin / discrete phenomenon.)

Humans have invented many ways to store data, *e.g.* knots in strings, marks cut on sticks, clay tablets, books, vinyl records, cassette tapes, optical disks, electrostatic memory, and so on. But every single one of these operates using the same principle, of dividing a continuum up into (somewhat arbitrary) discrete parts, then inserting "holes".

The universe originates in "one", which we can treat as a continuum between zero and one. A continuum provides the possibility for information to exist but contains no actual information until it's divided. It is the "aether" (substrate) of information.

If we place a "hole" somewhere in that continuum, we create information. We have divided the range into two discrete parts, and we now have two things which can be compared. We have created a "relationship".

So, simply by placing a notional "hole" (or a knot / not – interesting correspondence) in a "string", we create information, and this is the underlying form of all information.

- A continuum can be infinitely subdivided, and so can store a countable infinity of discrete units of information.
- Discrete items are always a relatively arbitrary subdivision of a continuum.
- The true infinity of Yang "continuum" is the "host" of the countable infinity of discrete Yin phenomena.

Consider this page of writing as an example. There is a blank, continuous page of white paper, arbitrarily subdivided into sections, filled with black "letters", *i.e.* specifically shaped shadows / holes.

Binary vs analogue

What about binary vs analogue data? While binary data is obviously full of zeros / holes, a vinyl record or cassette tape is an analogue recording, its data is continuous and not discrete.

Ultimately, all data are continuous / analogue. Discrete items are always a relatively arbitrary subdivision of continua. In other words, holes cannot truly be discrete. The edge of a lake always slopes.

All the ones and zeros in a binary storage are actually hills and troughs in a continuous medium. We arbitrarily limit the range of values it represents to just 1 and 0, but the medium itself can and does contain a range of values. Yang continua are real, Yin divisions are arbitrary.

The Paradox of Yin

The fundamental paradox within Yin is the concept of absence.
Absence "makes the world go round".
As we'll see later, in the section of physics, it is "the hole" which creates all curvature.

The "Lady of the Lack"

In nature, it's often the things that are missing which are the most important. Although the etymology is controversial, I suggest that a "lake" is a "lack", *i.e.* of land. It's the absence of land that makes a lake, and it's a good analogy. A lake is just a hole in the land, but when filled with water it provides a more productive habitat for life than flat land can. More niches, more biodiversity, more life.

Life prefers to live at the edges between Yin and Yang, land and water, earth and sky. It is at the boundaries between everything and nothing that we get all the interesting stuff.

lake
- Latin lacus ("lake, basin, tank"), see lac
- I suggest, related to lack, like, lacuna, lagoon
lacuna
- an empty space or a missing part; a gap

Note, you "like" things you "lack". If you like things, it's because you lack them and want more of them in your life. These ideas are two sides of the same coin. Again, these words are not "officially" linked, but probably should be.

Yin is the "lady of the lack (lake)". It is absence, omission, emptiness, void. It's "the thing that does not exist". If Yang is "something" then Yin is "nothing", and that is intrinsically a paradox. "Nothing" cannot exist, so how can we even talk about it?

Yin is the archetype of "nothing", false, paradox, illogic, and conflict. It is a paradox, but it can be resolved from a different perspective. Just as a dream "does not exist", but also does.

Reality is constructed in a hierarchy of multiple levels, and the UP says paradoxes are always an invalid comparison between levels, a category error. For example, Yin is "illusion" and "a dream", but if you're in a dream it is Yang to you, it's real. The dream becomes your "provider of reality", able to determine your fate.

Yin - Absence	Yang + Presence
To not be. Nothing. Non-existence.	To be. Everything. Existence.
The void / hole. Emptiness	Prominence. Fullness
Shadow	Light
Matter, solidity	Energy, activity
Temporary, temporal	Permanent, eternal

Examples

An example of the paradox within Yin is the concept of "knowledge", particularly of the "world" when the world is essentially an illusion. One enters the world (Yin) as Yang to know it, but it is really a projection of his own mind, he is knowing himself.

We can only know things by comparison with other things, but One is the only thing that really exists, there is nothing to compare it to. It is literally "incomparable" due to its aloneness. The only option is to compare it against its own reflection.

Yin is like a dream. In a dream you can see people, touch things, smell scents and so on, but when you wake up (and "ascend to a higher-level of consciousness"), you realise didn't touch anything other than your bed.

Here are some examples of paradox inherent in the Yin principle.

Paradox in measurement

To know the world we must measure it, but:

- Measurements are inaccurate, and units are arbitrary.
- Measurements convert reality from qualities to quantities and in so doing from absolute to relative, from objective to subjective, from the more to the less real.

Measurements are always approximations. It's not possible to measure any quality perfectly, there's always a margin of error. Results are always quantised. Continua are always arbitrarily divided.

Paradox in matter

Yin is "matter" and is fixed and solid, while Yang is activity, constant change, but:

- Matter is always changing and is in a state of constant flux.
- It's not solid, being made of intangible "activity".

Perhaps another paradox is that matter is "imperfect" and "full of holes", but that imperfection in form is perfect in function. So, matter is both imperfect and perfect, depending on perspective.

("Imperfections" in the physical elementary particles causes matter to form, charged particles have "holes" that desire to be filled.)

Paradox in language

Language is Yin to consciousness, it's a tool. Language exists because we are separate minds and need to communicate. We cannot share our experiences directly, we can only share them indirectly via a medium.

There is a hole *void* abyss between our minds and the purpose of language is to fill it. But we can only convey relative information in language, not absolute information such as experience ("qualia").

No words can explain "turquoise" to a person born blind.

Language exists to bridge the gap between two minds, but it can never truly fill it. We can explain our experiences to each other but cannot convey the actual experience.

Note that language is like the shadow. It can only convey a "2D outline" of what you intend to communicate. Words *labels are like specifically shaped shadows* holes, they cannot convey the full 3D reality. The receiver of the communication can only fill in those holes by personal experience / absolute knowledge.

When we hear a word for the first time, we create something like a new encyclopaedia page in our minds, but it starts out empty and must be filled in. It can be filled with two types of information, either first-or second-hand experience, *i.e.* knowledge or belief. Belief does not actually fill holes or provide knowledge though. At best it's a signpost to knowledge, at worst it is a roadblock to it. The only thing that can truly fill in the holes in your understanding is first-hand, direct experience.

If we understand the inherent contradictions in Yin we get a clearer picture of Yang, and that is the goal. While Yin may not deliver what it seems to promise, its imperfect delivery is what is required. It provides better contrast to the "big picture", it's more useful for comparison purposes, and it allows all paradoxes to be resolved. Imperfection is an inevitable and essential quality of Yin.

So, paradoxes exist within the UP, but without violating any rules of logic, and they contribute to the explanation it is designed to produce. It is an amazing system.

Chapter 4

Four Elements

The four elements describe the work of the universe. They tell us many interesting things about the nature of reality, including opening the door to an intriguing new theory of mind. So, where does this set of ideas come from, and why are there four?

Two Times Two

The simplest (and arguably only) way to create the next level of reality. Is to make a "duality of dualities", and two times two is four. This seems to follow necessarily from the existing structure.

To make the next level of entities, we must give them both a Yin and a Yang characteristic. Thus, each of the four categories has two properties, a Yin one and a Yang one. Each of those properties can be either Yin or Yang (-/+).

	Yin property	Yang property	Result / "Genome"
Category 1	Yang +	Yang +	+ +
Category 2	Yang +	Yin -	+ -
Category 3	Yin -	Yang +	- +
Category 4	Yin -	Yin -	- -

With this arrangement we can now have entities that are Yin in one way, and a Yang in another, so they're two-dimensional (2D). *I.e.* to specify one of these items, we must give two pieces of information. This allows things to have an "inside" and an "outside", like a circle drawn on paper has an inside and outside but a 1D line doesn't.

The identity, purpose, and properties of the four categories must be defined by this arrangement of Yins and Yangs somehow. Each one has its own unique combination. It's a bit like a simple version of DNA: each category has an "X" and a "Y" "chromosome", and a single "gene" on each chromosome which can be either "X" or "Y".

The big question is what are the four categories? They're a new "generation" of ideas, and must describe a new, more detailed level of the conceptual world.

A Process of Creation

If the purpose of the universe is to explain itself, then an explanation of the creation process should be embedded in the framework. We now have four archetypes that were used to create the universe, so they should somehow explain how it was created.

What we find is that they explain how to make things in general. They are the most generalised algorithm for making things that is conceivably possible, like a really vague "blueprint of creation". Let's consider the concept of creation itself. What things are common to all acts of creation / manufacture? What are the most basic concepts involved?

If all such acts are fundamentally the same at some level of detail, it shouldn't matter whether we're describing how to make a spaceship or a cup of tea. The process should always contain the same elements at that level of detail. We need to zoom out of the picture and look for the most generalised things that all acts of creation must have.

A note on sources

To be honest, I don't know if I would have figured out the link between the four elements and the process of creation on my own. When I began this research many years ago, I found some websites which linked the classical four elements to the conceptual steps required to create something, as described below. Unfortunately, I lost the links, and now I can't find them.

There are many websites out there about Alchemy, but most, if not all, seem to miss the mark these days. So, I must apologise for not being able to provide links to sources or related info, I just can't find any.

Four Elements

The underlying idea of the four alchemical elements is that there are four fundamental conceptual parts required in any act of creation, and these four archetypes describe a universal process by which all things are made. (I may refer to the four elements as "4E".)

The four things necessary for any act of creation are:

1. A Reason

In order for you to make something there must be a reason WHY.
You must have a desire or need to fulfil, a purpose.

2. Instructions / A Plan

There must be a plan, a list of instructions, a method / algorithm.
You need to know HOW you're going to achieve your aim.

3. Agency / Ability

You must have the ability to act and carry out the instructions (WHO).

4. A Possible Result / Reward

There must be at least the potential of gaining something you want from the effort.
WHAT is the reward?

Example, to make a cup of tea.

1. You have to want a cup of tea.
2. You have to know how to make tea.
3. You have to be able to make tea.
4. There has to be tea at the end of the process.

In other words, to do anything, you must have the will, a plan, and the ability. When you're done there has to be a result, a product, a reward. These four broad concepts will apply in every case of creation.

If we put these ideas into a list of synonyms we have:

1	Want, desire, motivation, will, intent, purpose, aim, spirit, beginning
2	Plan, design, map, algorithm, method, path, thought, information, knowledge
3	Ability, aptitude, skill, force, work, effort, strength, power
4	Product, reward, output, result, matter, end

The first three categories are all active, they are the inputs to the process, they all contribute to the production of the result, which is the passive output.

Are these the right fundamental concepts that should be in our four categories? They are all quite different, they do cover a wide range of ideas, so they might work. But even assuming they are the right ones, how can we assign them to the four +/- categories above? Which concept goes in which category, and why?

Well, we have their "genome" to study, their own particular combination of +/-. These describe combinations of properties, so they should predict the properties of their owners. In the table below, the two columns Yin and Yang are like the two chromosomes, the different combinations of genes possible are the 4 rows.

I'm using the properties of tangibility and activity to link the "genetics" of each category to the properties of the concepts on the right.

- The Yin aspect determines how physical (tangible) a thing is.
- The Yang aspect determines how "active" it is.

This is equivalent to "form / function". Active is equivalent to "function", inactive is equivalent to "form".

This table and the concepts it's trying to convey may take some time to digest. I explain the reasoning below.

		- Yin Aspect: Tangible PHYSICALITY	+ Yang Aspect: Form / Function ACTIVITY	Concept
Cat 1	+ +	Yang / Yang Intangible, formless, active, changing, functional		Will, desire
Cat 2	+ -	Yang Physical aspect is intangible. You can't touch information.	Yin Activity aspect is inactive. Information has form but is not a function.	Information, design, thought, path
Cat 3	- +	Yin Physical aspect is tangible. You can feel a force.	Yang Activity aspect is active / changeable. Force has function but no form.	Action, force, ability, work

Cat 4	- -	**Yin / Yin** Tangible, solid, inactive, unchanging, having form, passive	Product, result, reward

Category 1: "Desire" ++

The beginning of any act of creation is the desire for the result. Tea won't get made if no one wants tea.

- Yin property is Yang. Physical aspect is intangible, non-physical. You can't touch a desire / motivation.

- Yang property is also Yang. Its activity aspect is "active", meaning it has function (can do things) but has no form. You can't depict (make an image of) a desire, you can only depict the object of a desire.

A desire has no dimensions or shape, it's unquantified and vague, *e.g.* "I want to learn an instrument", or "I wish I could fly". Desires are "light" and can be changed easily. ("I've changed my mind, I'll have a coffee instead.")

Bear in mind this is a super-set of all concepts related to "desire". There are different types of desire, some are easier to change than others. Physical desires, the urges of the body, are the lowest, most solid form of desire, but even they come and go.

Category 2: "Thought" +-

For a desire to be manifested it must go through the "planning stage" to "solidify" it. Every act requires a plan of action. You can't make tea if you don't know how to make tea.

A plan is a sequence of fixed, solid ideas. They don't change as often as desires do. Cultural traditions are "thoughts" and can persist unchanged for many lifetimes.

- Yin property is Yang. Physical aspect is intangible. You can't touch a thought. A thought isn't made of matter, but it can be "mediated" by matter, just as news is mediated (communicated, transmitted etc.) by a newspaper.

- Yang property is Yin. Activity aspect is inactive, unchanging, fixed, solid.

A thought is information, and it has a kind of fixed shape / form. Your memories of the past are thoughts, and they have a "solid" fixed shape. Information "makes a form inside".

Category 3: "Work" -+

Once you know what the plan is you can do the work.

- Yin property is Yin. Physical aspect is tangible / solid. You can feel a force, it has physical reality, but like information, it isn't made of matter. Force is mediated by matter, but it's not strictly made of it.

- Yang property is Yang. Activity is active *changing* not-solid *formless* functional. A force has no shape because it's not made of matter, but it has function and can "do things".

Forces are made of intangible formless "energy", being transmitted by matter. The thing mediating a force might have a shape, but the force itself doesn't. A force does have a path of action though, a vector.

Category 4: "Matter" --

Matter is the product, the reward. It's the thing that was desired, made real.

- Yin property is Yin. Physical aspect is tangible / solid. You can touch matter.

- Yang property is Yin. Activity aspect is inactive / fixed *unchanging* solid and has form.

Classical Elements

The classical four elements are described in Wikipedia as a theory of everything, but one that isn't supported by science.

Classical elements typically refer to water, earth, fire, air, and (later) aether, which were proposed to explain the nature and complexity of all matter in terms of simpler substances.
Modern science does not support the classical elements as the material basis of the physical world

`https://en.wikipedia.org/wiki/Classical_element`

The article describes them as a proposed material basis for reality, but this is a misconception. The classical four elements are not material things but archetypes. They were never intended to be thought of as a material basis for a physical universe, but as a conceptual basis for a non-physical one.

It's easy to see how, from the materialist perspective, the 4E seems to be an irrational idea. A physical element like gold can't be made of a molecule like water; that makes no sense. The 4E only make sense in the context of an idealist view of nature where they can be viewed as something like "Platonic forms".

Aristotle's Elements

This book isn't intended to look at all the different views on the topic over history, but it's necessary to explain how the Aristotelian scheme is different from what is being described here to avoid confusion.

The Wikipedia page above tells us that Aristotle assigned these two properties to the elements: hot *cold, wet* dry. If we assume that Aristotle was using dryness and temperature as symbols for Yin and Yang, does his system concur with the relationships we've already established?

	Dry/Wet (Yin Aspect)	Hot/Cold (Yang Aspect)	
Fire	Dry = Yang	Hot = Yang	+ +
Air	Wet = Yin	Hot = Yang	- +
Water	Wet = Yin	Cold = Yin	- -
Earth	Dry = Yang	Cold = Yin	+ -

Aristotle's scheme doesn't agree with the one we're considering because he describes water as "cold and wet" which are both Yin characteristics, so the systems don't match. Water and Earth are reversed in Aristotle's depiction.

The Aristotelian scheme lacks a firm connection to duality which is where the properties of the 4E originate. The system described by the UP is much clearer and more substantial.

Alchemical Elements

Instead of viewing them literally, the 4E should be viewed as a template or algorithm. More like a conceptual tool, or a sequence of categories or instructions. They are somewhat hard to define because they are highly generalised super-concepts containing a diverse set of ideas.

What is the significance of the symbols of Fire and Water etc? Why did the alchemists of the past use those substances as the category names?

Four States of Matter

The 4E correspond to the four basic states of matter (amongst other things), so that's probably a major part of the reasoning. The basic states of matter are:

- **Fire** is a form of plasma, an ionised gas capable of carrying an electric current.

- **Air** is a gas. Gas is relatively light and compressible. Its molecules "repel" each other, their interactions are straight / direct.

- **Water** is liquid. It's relatively heavy and incompressible but not solid. Its molecules attract each other but can still move. Their interactions are circular / rotational.

- **Earth** is the solid state, made of molecules that are stuck together and don't move.

This natural ordering of the states of matter is F A W E, as listed above, and this is significant. We get from one element to the next by cooling down the substance, or heating it, and they naturally appear in this sequence. This is the same order as they appear in the human body.

"Fire" Transforms

Plasma is the least dense, lightest form of matter, and it's made of charged particles / ions. Plasma can emit light, electric and magnetic fields, it's more "active" than the other states, and can exhibit a wide range of complex "behaviours" (e.g. lightening. Northern lights, flames, ropes, corona discharge).

Charged particles experience strong forces of attraction and repulsion. We can equate these to "desires". Electrons "desire" to be apart, unless in parallel motion, in which case their magnetic field attract. It's this balance between repulsion and attraction when in motion that makes the behaviour of plasma so complex and interesting.

"Fire" is the archetype of motivation / will / desire. We experience "burning desires". Both fire and will transform things from one form into another, and both can reproduce themselves.

Fire transforms tangible matter into intangible gas, light, and heat. It changes the lower element of Earth into the higher elements of Air, and more Fire. It can reproduce, like a living creature.

All the work / transformation of the world is driven by people's desires. Desire can transform solid matter into all kinds of things, including children. Desires can also reproduce by moving into new minds. *E.g.* the desire to watch a particular film because your friend recommended it.

"Air" Informs

While Fire is desire, "Air" qualifies and quantifies it. Air defines the path we will take to achieve our desire.

Air (the plan) is the "product" of Fire (desire), just as Earth (the result) is the product of Water (work).

Plans are will in a crystallised form that can be communicated to others. Air is information, the intangible form of matter, it is "solidified desire". Fire is the motivation; Air is the plan to satisfy it. The Air element is closely related to logic, and contains thought, intellect, communication, decisions, and "laws".

law
- a system cataloguing standards and comparisons intended to act as a guide for action

Air is the archetype of information, the way, the design, and the law. It informs but doesn't compel. It indicates the right direction, it's a "guide". You can have a plan, but then change it because of circumstances. You can have information but ignore it.

Air must be considered in comparison to the Water element because they're mirror images of each other; while water can carry a lot of force, the wind's not usually that strong. Air is light, whereas water is heavy.

Water is visible, but air is invisible. The wind is like thought, we can't see the wind, but we can see its effects, *e.g.* the leaves on the trees move. We determine how strong the wind is by looking at its effects. The same is true of laws *rules* plans / thoughts. We can discover the laws a thing operates under by observing its actions and their results.

We express our thoughts via the physical medium of air, by our speech. Speech is our main method of communication / transferring information, our thoughts are "expressed in air" (via the Voice mechanism).

"Water" Is Force

Water is heavy, it can carry a lot of force. It has no defined form, so it can flow to fit any shape; it's ultimately flexible and adaptable. Air is soft and can be compressed, but water is hard and incompressible. The Water element is the archetype of all concepts related to power, force, work, action. Water can sculpt landscapes by erosion or tidal wave, it can break mountains into sand.

Work is the principle that can convert an intangible idea into a tangible object. Work is "Alchemy", it's the process of "transmuting elements". I would suggest that phrase was not primarily intended to apply to the physical elements like gold, but to the four conceptual elements: Water (Work) converts the Fire (Will) and Air (Design) elements into the Earth element (Product).

Also, water can carry waves on its surface, and this category correlates with the principle of the sinewave. It defines all circular phenomena, as we shall see.

"Earth" Matters

Work creates products. Products are what people desire. If the universe was created to fulfil a desire, and it's full of matter, then matter was the object of desire. It is what was intended. The Earth element is the archetype of the result of all our planning and effort. It's solid, tangible, and three-dimensional. It's the object of our desire made real, it's "wealth".

Matter is what matters. It's what all the effort is made for. We work to obtain, optimise, and maintain things made of matter, the house, the car, the family *etc*.

The Human Body

How does the set of the four elements manifest in the material world? How can we personally observe it? If it's a "universal form" it should be everywhere. The human body is a material thing, it has a fully defined 3D form. If we were to find that it is arranged in the order FAWE, that could be significant.

Funnily enough, the human body is arranged in this order from the top-down.

Element	Body Part
Fire State: plasma = electricity Will, desire, intent	**Brain** Runs on electricity Seat of consciousness, will.

Air State: gas Law, plan, rules, decisions	**Lungs** Runs on air We communicate our choices using air
Water State: liquid Force, power, activity, strength, ability	**Stomach** Runs on water (mainly) Stomach is source of your strength / health?
Earth State: solid Matter, solid objects, product	**Limbs** Legs run on the earth (pun intended) Limbs interface with physical matter.

The human body is indeed arranged in the FAWE order of the 4E, with Fire at the head, and Earth at the feet. The parts fit well, with the brain running on electricity, lungs on air, and so on.

It is arranged in this order from the outside in.

Element	Body Part
Fire	Nervous System - Electrical Most nerves are on the outside, in skin.
Air	Circulatory System - Providing Air Blood vessels are mostly on the outside.
Water	Muscles - Work Muscles contain a lot of water.
Earth	Bones - Support Bones are the most solid part and are the foundation of the body.

Other Fours

The elements correspond to the arrangement of our environment.

Element	State	Environment Part
Fire	Electric Plasma	Sun
Air	Gas	Sky
Water	Liquid	Rivers, Seas, Clouds
Earth	Solid	Land / The Earth

For clarity: there are several fours that exist in nature, most of them do not seem to categorise naturally or unambiguously into the four elements. For example:

- **Four cardinal directions.**

How these could be categorised in a satisfactorily objective way, I'm not currently sure.

- **Four dimensions.**

Einstein's "spacetime" has four dimensions, but this theory doesn't see the relationship between space and time in that way.

- **Four basic math operations.**

Addition, subtraction, multiplication, division.

These maths operations do categorise into the UP, but not as the 4E.

- **Four fundamental forces.**

Modern physics believes there are four fundamental forces, but this theory suggests there is only one, although it does manifest in multiple ways.

Also.
- Four nucleotides (DNA).
- Four types of teeth.

We do however find clear links between the 4E and:
- The human body layout (and the heart).
- The states of matter.
- Computing / information management.

- The "process of creation".
- The sinewave / seasons.
- The structure of language.

The Computer Analogy

One claim of this theory is that the universe can be considered as a computer. It is either analogous to a computer or is actually one. Being (like) a computer, it has the same parts as any human-made computer has, and we can usefully describe it in those terms and draw analogies.

The 4E can be viewed as a universal computer. Together, they form the simplest-conceivable system for processing data. They are the archetype of "work", and processing data is the work of an information-based universe.

A computer can be thought of as having four basic levels, and these could be correlated to a simulated universe as follows.

Element	Computer	Universe	One to many
Fire	Hardware	Consciousness	One computer
Air	Operating System	Concepts / the UP	One computer to many OS's
Water	Program	People / creatures	One OS to many programs
Earth	Data	Experiences (matter)	One program to many datasets

In this analogy:

Fire - hardware / consciousness

Consciousness, the "Universal Mind" is equivalent to a computer's hardware. It's a singular, unchanging thing. It's the "container", "host" or "substrate"; everything below is contained within it. Fire is the archetype of plasma / electricity, and brains and computers run on electricity.

Air - O/S / the UP

The "Operating-System" of the universe is the UP, the conceptual-framework. It provides and defines all the "services" programs can use and how they work.

It's the "Law" element. This is like the "spirit world", where the "gods of nature" would be equivalent to OS services (like disk-management or user-management). These processes are sometimes called "daemons" in computing.

Water - programs / people

The purpose of a computer is to run programs because they perform functions which are useful. They exist on top of the OS layer and don't need any knowledge of the hardware level to work. Just as humans can operate without understanding duality.

We can (broadly) view both software and people as "filters" allowing a narrow, specific view of a dataset. Programs provide filtered views of data. Individual lifeforms have specific views of reality.

Filters create different views / perspectives of a set of data, they reduce a large set of data into a smaller, more manageable form that can be acted on. For example, consider software like YouTube, or eBay. They have a vast dataset, which you access via various filters.

Programs are useful because they help us deal with specific, limited datasets. Human perspectives are valuable (to "the universe") because they do the same thing. We are all unique, and we're unique because we are limited and specific.

Earth - data / experiences

Individual items (or sets) of data in a computer program equate to individual "experiences" and are the "matter" to be processed. Data is the non-material equivalent of physical matter. Programs process data, and so do people. We run on different types of data though. Computers can only handle quantitative data, whereas minds mainly process qualitative data.

While the data types are different, the principle is the same. The dataset a program is processing at any one moment is analogous to an individual experience you have. An individual experience is like a frame in a movie, or a particular state of a program; all of which can be considered "just data".

In this analogy, you and I are like individual instances of the "human" application "running on" the hardware that is the universe. Every moment / experience we have is like a new dataset being processed.

Ultimately, this arrangement suggests a foundation for a computational theory of mind. It suggests that all our experiences can be generated procedurally from a set of logical rules, just as data can be. The underlying structure of mind is identical to that of a computer, although there is one important difference we'll come back to.

The Turing Machine

The four elements align with the four basic parts that are necessary for Turing-complete computation. There are plenty of good sources on Turing machines on the internet, *e.g.* Wikipedia.

`https://en.wikipedia.org/wiki/Turing_machine`

A Turing machine is the simplest universal computer that can do any calculation, and its instructions need four parts to provide this functionality. This table shows the four components / stages of an instruction.

Part	Description	Machine Instruction
Earth	Present location / situation Two-dimensional, location in a 2D table. Where am I?	INPUT: Current state + input symbol
Water	Action What should I do here?	OUTPUT 1: Write symbol
Air	Plan / Path Where do I go next?	OUTPUT 2: Move tape left / right
Fire	New "State of Mind" / Motivation ("Will") What do I "want"?	OUTPUT 3: Next state

A program for a Turing machine is written as a table, a 2D matrix where "current state" and "input symbol" are the X, Y coordinates of an individual table cell / instruction.

The 4E are the 2D level of the system, they state that all forces are 2D; that all work is done using 2D forces. The Turing machine is a 2D device, both its input program and output result are 2D, and can be described in a table.

Properties are 1D, work is 2D, matter is 3D.

EWAF

The ordering of the 4E here is the reverse of what we have seen so far, it's EWAF instead of FAWE. Why is that?

The direction of the elements in the table above is from "Earth" to "Fire" because it's going from matter (location *raw data*) to will (*instruction* decision).

We start out with a location and end up with a new "motivation" / state. This process is the archetype of "making a decision", and as we'll see is also the archetype of work in general.

Note, the input consists of two items of data, but they both belong in the Earth category; Yin is many, and they're both the passive result of previous operations, so this fits that archetype.

One-Instruction Set Computer

"an abstract machine that uses only one instruction... an OISC is capable of being a universal computer in the same manner as traditional computers that have multiple instructions"

`https://en.wikipedia.org/wiki/One-instruction_set_computer`

This is a class of computers that people have explored to try to understand the limits of computing. It's purely for fun, they're not practical, but again it shows we need a minimum of four things to do computing.

Two examples of the type of single-instruction you can use:

- Subtract and branch if negative.
- Subtract if positive else branch.

The four parts are as follows.

Stage	Description	Part
Earth	Location / Instruction	Current Position, Instruction Pointer
Water	Action	Subtract A from B
Air	Plan / Law / Rule	Branch if B is Negative
Fire	New State / Motivation	Branch To

Relational Databases

We also need four elements for a fully functional database.

Relational databases are designed to operate using the logic of sets. In a relational database you have "tables" of data. A "table" is the same thing conceptually a set, and it contains a list of things. *E.g.* we could have a database table called "Shopping List".

One database can contain many tables. One table can contain many columns, and many rows. There are four basic levels to a database, and they increase in detail and the number of dimensions as we descend through them, they fit like so.

Element	Dims	Level	Description
Fire (Purpose)	0	Database	A single database. *E.g.* "Bob's Carpets Ltd"
Air (Plan)	1	Tables	A list of categories of data. *E.g.* Products, Stock, Customers, Sales, Orders ...
Water (Work)	2	Columns	Each table contains a list of columns (sub-categories). *E.g.* in the table "Products", the columns might be: Product ID, Name, Description, Category, Price
Earth (Product)	3	Rows	Each row contains the actual data / information about things. *E.g.* an example product might look like the table below.

The top three elements describe the structure the data must take, the Earth element contains the actual data. An individual database table with data in it can be represented as a 2D table, *e.g.*

Table: Products					
Row ID	SKU	Name	Description	Category	Price
1	CBC001	Blue 001	Blue loop weave	A	11.95
2	CBC002	Blue 0022	Blue cut pile	B	12.95
...any number of rows...					

To address any individual item of data within a normal relational database you need three items of data, *i.e.* three dimensions. (The database is the container; it doesn't need to be specified in queries.)

1. Table name
2. Column name
3. Row ID (or other selection criteria)

In the computer analogy "data" is the Earth element, so is equivalent to physical matter even though it's intangible. It's a 3D non-physical phenomenon. An instruction to fetch an item of data could look something like:

```
SELECT column FROM table WHERE row_id=1
```

We'll see more real-world examples of the 4E in due course.

Arrangements of Four

Four different things can be arranged in 24 different permutations (4!, four factorial), although half of the set is mirror images of the other half, so there are really 12 unique arrangements. However, so far, I've only encountered the four elements arranged in two different ways.

Different arrangements of the 4E constitute different "compound concepts", like "sentences", or "plans". The individual elements are like words, so a sequence of them is like a sentence.

Any combination of the 4E in sequence implicitly includes the concept of stages / phases. They will necessarily be time-ordered concepts. We've already encountered the order: FAWE, and EWAF, it's reverse. This appears to be a single pattern / process that can run forward or backwards.

The four also define the form of the sinewave *circle* seasons. In this case they appear in the order WFAE. This pattern might also be reversible. These two patterns seem to have opposite properties, they are probably two halves of a duality.

These two patterns seem to neatly arrange into "function" and "form". However, more correspondences are needed to confirm this description, it's almost too perfect. I don't know if any other possible patterns of the 4E occur in nature, these are still early days for the theory.

FAWE: "Universal Form"

This pattern defines how things are created, or at least how it looks on the outside. It seems to define the "form of creation", or something like that. The direction Fire to Earth describes a process that transforms desire into a product, like making tea, or a universe. It describes will being transformed, via work, into matter.

Stage	Description	Creation	Body	Database	Matter
1. Fire	Desire	I want a drink	Head	Database	Plasma
2. Air	Plan	"I will make a cup of tea"	Lungs	Table	Gas
3. Water	Work	I make some tea	Gut	Column	Liquid
4. Earth	Result	I now have some tea	Legs	Row	Solid

EWAF: "Universal Function"

The reverse of the pattern above is EWAF.

Earth to Fire describes a process that "transforms matter into will", *e.g.* making a decision prompted by events in the environment. When you decide, you create a new desire (direction) based on material facts. It has the form of being the reverse of the creation process described above.

To make a decision, the facts of the matter must be considered and converted via a choice into a will / intent. It's the same form as an instruction for the universal computer we saw previously. It also turns out to be the pattern for work in general, so EWAF is the "universal form" of work and choice.

Stage	Description	Part	Battery / Work
1. Earth	Data / facts	INPUT: Gather the information you need	Discharged Spade empty
2. Water	Order Data	PROCESS: Filter / prioritise the facts.	Charging Fill spade with dirt
3. Air	Plan / Law / Rule	CHOOSE: Decide which is the logical course of action	Charged Move spade to output area
4. Fire	New State / Motivation	OUTPUT: A new motivation *state* direction	Discharging Empty spade

The output of this process is an intangible desire. For example, if after some deliberation you decide to sell your car, then it's true to say you "want to" sell your car, it's a desire. The effort of your deliberation resulted in an immaterial thing.

A Linear Cycle

In real life our acts of creation are usually prompted by the facts of the outside world. We do things in response to problems or situations we encounter. We can combine the two lists above and traverse it going up and down.

Stage	Description	Example
Earth	Situation, Problem	I am uncomfortable
Water	Discernment, Work	What is this feeling? (Search, filter, compare)
Air	Law, Judgement	I am thirsty.
Fire	Desire	I want a drink
Air	Law, Plan	I will make a cup of tea
Water	Physical Work	I make some tea
Earth	Result, Solution	I now have some tea.

So, this pattern can be used in a kind of cyclic way, going from Earth to Fire and back again. The motion is a bit like a boat floating on waves, "bobbing up and down" through the sequence in a linear cycle. It's not a true cycle in the sense that it doesn't form a circle, the ends aren't connected. This pair of patterns is perhaps describing the perspective of "bobbing up and down" on the wave WFAE.

WFAE: "Circles"

The 4E also define the concept of the sinewave *circle* seasons, in the order WFAE. This ordering allows them to join up in a loop while only changing one "gene" (+/-) at a time. I.e.:

FAWE	+ +	+ -	- +	- -
WFAE	- +	+ +	+ -	- -

To discover how the 4E describe a wave, or any 2D process, we need to find the two fundamental properties of the phenomenon and then link them to Yin and Yang in a table of the four possibilities. Classification is quite simple in the case of the wave, the properties which constitute it are just "position" and "direction".

The wave process is created from this complementary duality of opposing dualities. Position / direction as a pair is complementary, but individually they are opposing. *i.e.* position breaks down to "high / low", and direction to "up down". *(Note: There are also 2D objects made of a complementary duality of complementary dualities. We saw this when looking at "inside outside".)*

Position: Is it above zero?

The position of any point on the wave can be either positive, above the (zero) centreline, or negative, below it (high/low). In a compression wave this corresponds to pressure being above or below the ambient value. In the seasons this represents a day-length of over or under 12 hours.

Direction: Is it increasing or decreasing?

This can be either positive, towards the top of the graph *wave, or negative, towards the bottom (up*down). In the seasons, this represents whether the day-length is increasing or decreasing. In a compression wave it would be whether the pressure is increasing or decreasing.

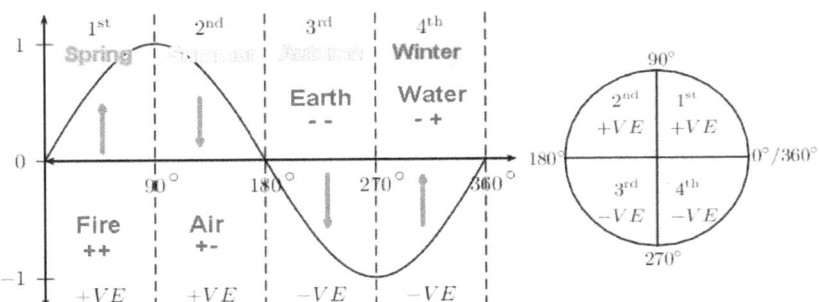

Position and direction form a complementary duality of properties that describe a wave. Position is the passive result of direction which is the active part, so position is Yin, and direction is Yang. This is everything we need to classify it.

In Spring, the amount of sunlight is above the zero-line (> 12 hours), so its position (Yin) is positive (Yang). Its direction (Yang) is also Yang because the length of the day is increasing. It corresponds to Fire as both its Yin and Yang properties are Yang / positive.

In Summer, position is positive as day length is over 12 hours, but the direction is negative, the days are getting shorter. Hopefully you can see how that works for the other quadrants of the graph / seasons.

Season	Element	Position (Daylight over 12 hrs?)	Direction (Day getting longer?)

Spring	Fire ++	+ >12	+ Longer
Summer	Air +-	+ >12	- Shorter
Autumn	Earth --	- <12	- Shorter
Winter	Water -+	- <12	+ Longer

Below is included the concept of life cycle of a human from gestation to death which indicates these cycles are analogous. We'll revisit this and add more detail in a moment.

Season	Reason
Spring / Fire ++ Desire Adolescence	Fire is desire. Spring is the time when all living things are full of desire to mate. Seeds are planted. The direction for the rest of the year is set.
Summer / Air +- The Path Adulthood	Air is the way *path* plan. Everything is following the path set in spring. Crops and babies are growing. (Yang *hot on the outside,* Yin *passive on the inside*)
Autumn / Earth -- Decline and Death Old Age	Autumn is the season of death, when all fruit is harvested, and everything returns to the earth. (Hence Halloween / Day of the Dead in its middle)
Winter / Water -+ Gestation and Rebirth Childhood	Winter is preparation for Spring. New buds and desires are formed. Water is action and Winter corresponds to childhood, the most active time. (Yin / cold on the outside, Yang / active on the inside)

It gives us the pattern FAEW, but that's starting the year in Spring. Starting in Winter gives us WFAE.

It repeats: ...FAEWFAEWFAEWFAEW...

As another example, this table shows the phases of a pendulum. It's the same as the seasons / sinewave, it just has the property of left/right instead of up/down. The pendulum is also analogous to an oscillating electric circuit ("tank circuit" *LC oscillator*), *where left* right is equivalent to - / + charge.

	Pendulum / Oscillator			
Phase	Position	Direction	Description	
Fire ++ Spring	Right +	Right +	Pendulum is to the right of the centreline, and it's going right.	
Air +- Summer	Right +	Left -	Right of centre, going left	
Earth -- Autumn	Left -	Left -	Left of centre, going left	
Water -+ Winter	Left -	Right +	Left of centre going right	

Note, the name "winter" is etymologically derived from the word "water", so there is a linguistic link to the associated element, at least for that season.

https://www.etymonline.com/word/winter

The four seasons of the year begin at (are demarcated by) four "events", the two solstices and equinoxes. If we add those into the list, we get 8 entities.

A Human Life

We can correlate this pattern of eight to the stages of a human life which does two things: it gives us an objective starting point for the cycle, and it also suggests that life / death is a wave-like phenomenon.

The cycle natural begins and ends at the winter solstice, the lowest point of the wave, and correlates to both the death / birth stages of life. It's the death of the old year, and the birth of the new one.

The "events" where the elements mix are like active "transformation-points" which set the path for the following season. These four mixtures of elements explain and fit the events in much more detail than is immediately obvious, or that we can cover here.

Element	Season / Event	Stage of Life
Water	Winter	Childhood
W + F	Spring Equinox	Puberty W + F = Action + Desire = "Desire comes into

		action / begins"
Fire	Spring	Young Adult
F + A	Summer Solstice	Becoming A Parent F + A = Desire + Law = "Desire to do what is right (i.e. for family)"
Air	Summer	Raising Family
A + E	Autumn Equinox	Family Have Grown Up A + E = Law + Product = "The path is complete"
Earth	Autumn	Old Age
E + W	Winter Solstice	Death / Rebirth E + W = Matter + Action = "Bringing matter to life (into motion)"

Just Two Patterns?

I don't know if these two patterns are the only ones that exist. I rather expected there to be (at least) four. They do seem to be a duality, but at this stage I'd like to keep an open mind.

The FAWE/EWAF pattern does "bob up and down" in a wave-like way. The question might be whether this is the cause of the wave pattern WFAE, or vice versa?

Given that there are 12 possible combinations of the 4E, we might expect to be able to correlate them to the 12 signs of the zodiac, potentially explaining their origin. Although, if they don't occur in nature, I'm not sure it'd be useful.

This is an open topic.

Four Senses

The UP says there are four basic categories of physical sense, but every category in it is associated with some kind of sense. We have internal and external senses, and some which are a mixture of both. It's a little complicated, and we'll come back to it later in the section on "Parts of Mind" to add more detail.

For now, we'll just consider the concept generally, in the context of the four elements.

Senses as "gods"

Our senses provide us with information about our environment. They are the mediators between (external) reality and consciousness, sitting in-between matter and mind. It can't really be overstated how important our senses are to us both practically and conceptually. Our senses are "gatekeepers" to the physical world, they are the only evidence we have that it exists.

It is conceivable that our physical senses are "fake". If this world is a simulation, then our senses are provided by the host machine and are conveying a fiction. It would, in theory, be impossible to distinguish if this is the case. The possibility this reality is something like the "The Matrix" (movie) can't logically be discounted, and we'll cover the full reasoning later.

Our senses provide / transmit information to us, we receive it and react to it, so they're Yang to us. Our senses are "like gods" to us, providing us with the arguably the most important feature of existence, "experience".

It seems we are mere receivers, dependent on our senses like a radio depends on a transmitter, at least according to this classification, at this level of the hierarchy. We don't know if our senses are telling us the truth, but we must assume they are for now, we have nothing else to go on.

Conscious / experience

Without "experience", can there even be "consciousness"? What is the relationship between these ideas?

I would suggest that consciousness / experience is a complementary duality. Consciousness is Yang: it can exist, but can't be described, without the latter.

Four physical senses.

Our senses match the pattern of the 4E very well. Modern culture says we have five senses, but the more Yin senses come in multiple forms because "Yin is many". The most Yin sense is the most material one, touch, which has the most different forms including pressure, pain, temperature, stretch *etc.*

The UP says we should categorise both smell and taste as two aspects of the same sense because they work by the same mechanism (solution).

We also have both internal, and external senses. The internal ones aren't generally recognised as such, let alone as having a direct relationship to the external ones; but I think there is significant correspondence.

Our internal senses are desire, judgement, emotion and memory. Hopefully you'll see what I mean.

State	External Sense	Medium		Internal Sense
Fire	Plasma / Electricity	Vision	Light / EM Wave	Desire / Will
Air	Gas	Hearing	Air	Judgement, "law" (good/bad)
Water	Liquid	Smell / Taste	Water (Solutions)	Emotion, "feelings"
Earth	Solid	Touch / Pain etc…	Matter	Memory

Inner Senses

The external senses work with things in the outside world, as opposed to the internal senses which work with the desires, knowledge, and emotions that exist on the inside. We need both sets to function. Correlating them via the UP hopefully brings a more balanced view to the concept of "sense" in general.

"Mind" is the set of (ten) internal senses

The "inner senses" listed above obviously correlate to various parts of "mind".

The UP tells us that the mind is a type of machine (we'll get to a full definition later). It is a set of cooperating components designed to do work. The work of the mind is to make comparisons, measurements, and decisions.

We can usefully view the phenomenon of "mind" as a set of senses, and as an information processing device.

The mind is constructed from all ten parts of the UP. Each category describes a different aspect of it.

The purpose of the inner senses is to "compute".

To understand the inner-senses, and why they're arranged in this order, think of it as the sequence of things you would do when you encounter a new thing, for example in a primitive survival situation ("the jungle"). This sequence goes from Earth to Fire because it's the process of forming a decision.

	Internal Sense	Description
Earth	Memory / Recognition	"What is it?" When you see a thing, this is the first question. Your memory allows you to "recognise" things. e.g. "It's a tiger"
Water	Emotion	"How do I feel about it?" What emotions does the thing elicit, fear or love? Emotions are powerful motivators. e.g. "I'm terrified."
Air	Judgement / Plan	"What should I do?" Consider options. e.g. "Don't move, I don't think it's seen me."
Fire	Will / Intent	e.g. "Be as still as possible, and hope it goes away."

There are also internal senses which correspond to the One, and Yin/Yang categories. The "One-sense" naturally corresponds with our sense of "identity". Duality, I suggest, provides the sense of "correctness" or "intuition".

- One: the sense of identity: consciousness, being, self.
- Yin *Yang: conscience, correctness, right*wrong, true/false, "intuition", **morality.**

There are also senses associated with the three operators (Voice, Heart, Sex), but we haven't got there yet.

Inner vs Outer Senses

How do the inner and outer senses correlate?

Fire: Vision / Desire

Vision corresponds reasonably well to "desire"; it is used to seek out the things you desire, but so are the other senses albeit to a lesser degree. In nature a hunter usually uses vision to locate prey, which is the object of their desire.

There's a saying: "what the eye sees the heart desires".

Air: Hearing / Judgement

In human and animal societies individuals (often) communicate via speech/sound. We tell each other what we want by talking (although sometimes simply grunting will do). If we have a plan, made a choice or a judgement, we tell people about it with sound. So, plans and choices are primarily conveyed by the air.

One thing I've noticed is people tend to sigh when making a decision, or when coming to terms with something. Lungs are Air, and certain breathing patterns seem associated with the process of decision-making / judgement.

Water: Smell and Taste / Emotion

The Water element is aligned with the principle of work *action* power, and with emotion. Smells can be very powerful and emotional. They can re-connect you with emotive memories long forgotten.

Scents are powerful motivators and both animals and plants use them to invite or deter others. Pheromones can attract mates from miles around, flowers can attract pollinators, and skunks can repel predators.

There's no doubt that smell is a powerful sense, it can elicit strong emotions including sexual desire / "love", perhaps the strongest motivator of all. Smells (and tastes) can cause significant changes in behaviour.

Earth: Touch / Memory

The Earth category includes the concept of solidity, a firm base to build on, and "products", things you have created which have a fixed form, like memories. Your memory is your "body of knowledge", and all internal activity is "built on" that. It acts as a foundation for all the internal senses above. Memory is the foundation of thought.

We could define memories as "a record of the shapes of things that have touched you".

Fooling The Senses

The external senses are Yang to the internal senses because they provide them with the raw data about the environment. This means the external senses have all the associated Yang characteristics in comparison to the internal senses. They must be more realistic and accurate than the internal ones which depend on them. *E.g.* if you see an elephant with your eyes, it'll be more accurate than one you recall from memory.

Optical illusions show how our "senses can be tricked", but this classification suggests the problem isn't the data received from the eyes, it's how it's processed. It's not the hardware, it's the software that's "glitchy".

The optical, auditory, and other illusions we experience are due to how the internal senses interpret the data they're given. Some might argue that the external senses don't provide accurate data about the outside world, but this model suggests they are relatively accurate.

To use another computer analogy, senses are like the computer's input peripherals, keyboard, mouse, scanner *etc.* Their input is accurate, but the software can misinterpret it, and this is a principle, as in the duality offer / choice:

Yang offers, Yin chooses.

The external senses just send the data they receive faithfully. They don't interpret it; they simply provide it. They're direct and simple. The internal senses are the ones that can get things wrong. They filter and interpret the data and make the choices, they're more complex and so naturally more prone to error.

Simulated Senses

Our physical senses could of course be simulated in a virtual reality, but could the internal senses also be "faked? Could our thoughts and feelings be generated as part of the "game"?

The lower two (mortal) "Earth" elements of memory and emotion would certainly be simulated. Emotions are obviously generated by the body, and if the world is a simulation, then our memories are too.

The upper (immortal) "Heaven" elements of law and will could also be faked / scripted, but if reincarnation exists and/or this is a simulation, at least some aspect of them must persist, and could come from outside. Most of our desires and plans are "worldly", *i.e.* to do with the "game", but we can have "higher desires". There is a duality there we'll explore shortly, the "Yin and Yang minds".

The UP implies that the (fundamental) Fire and Air parts of us are "immortal" and persist between incarnations. It also says those parts are the (only) ones which are "programmable" by us, we can change them and (some parts of) those changes are permanent. These "immortal" parts of us are the only ones which are truly "personal". They tell us about ourselves, we can change them, and they could conceivably persist between incarnations as a "personality".

Personal / impersonal

Most of our senses are "impersonal", by which I mean they come from "outside" us. We cannot change them as they are telling us something about the outside world / body / UP.

However, at least some aspects of our desire and judgement senses (Fire and Air) are truly personal. We "own" them and they tell us about ourselves, *i.e.* what we desire and think.

We can adapt how these senses operate by changing our "filter code", and that can fundamentally change how we see the world in general. By adapting our "desire and judgment filters" we can perceive the universe in a wholly different way.

So, these are our "personal senses". They belong to us; we create them by our own choices. My personal sense of judgement has changed dramatically over the course of life, as have my (physical / "worldly") desires. Although, I perceive this as having been driven by an underlying non-physical desire, essentially a "learning imperative" which hasn't ever changed.

Sense of identity

The UP implies we cannot change our sense of identity, which corresponds to the One category.

The archetypes suggest it's an "absolute" perspective and, in a sense, "impersonal" (somewhat ironically). We are not free to modify it, it doesn't really "belong" to us, it is "lent". Just as the dreamer "lends" his sense of identity to his dream-character.

The sense of identity *being* consciousness is a universal principle and so it probably feels the same to all living things. There is only one sense of identity, and it's universally identical. While the subjective experience of being a beetle will be very different from being a human, the sense of "me" would be the same. The beetle would still feel like "me", an individual observer experiencing the world.

The sense of identity unites all living things. It feels the same to a bacterium as it does to God himself, at least that is the implication.

Conscience and instinct

The sense of "correctness" (duality) is also not something we can change. Again, it is a universal principle that presumably feels the same to all living things and delivers equivalent content.

"Conscience" comes in two forms. It is an expression of the Yin and Yang minds.

The Yin form would be more like "survival instinct" (amorality), the Yang more like "social instinct" (morality). The purpose of the Yin "conscience" would be survival in the context of being alone, the purpose of the Yang is survival in the context of being in a group.

Both survival instinct and "morality" are impersonal senses. They're absolutes, defined by natural principles outside our control. Both can be understood as products of nature and evolution, and their content is the same for everyone.

Morality can be explained as a product of group dynamics, without needing to refer to any supernatural phenomena (i.e. the One category). Our "moral conscience" is a deep-seated sense necessary for social interaction, which is hard-wired into brains by evolution*. This is covered in more detail in the section on "Morality".

Notes:
- We'll come back to the senses again in the section on "Parts of Mind".
- We'll come back to the concept of our senses being simulated in the section on "Free Will".
- * The UP is compatible with the general concept of biological evolution, but with some differences.

Self-Categorisation

The levels of reality we've been exploring so far correspond to the first three of the four elements.

The four elements provide a more detailed classification of what unity and duality are by comparing them against each other in different ways. This means unity and duality fit into the classification of the four, they are described in more detail by them. Unity is Fire, duality is Air. The 4E classifies itself as the Water element.

Stage	Description	Dimensions
Fire - Will	Unity, One	Zero dimensions.
Air - Law	Duality, Two	1D objects: Properties
Water - Action *	The Four Elements	2D objects: Forces
Earth - Result	The next level...	3D objects: Matter

* We are here, currently discussing the "Water" level of reality. The next level must be "Earth". The 4E tells us there's only one more level of detail to describe. We're almost there.

Chapter 5

Seven Principles

Earth is the final element in the list, it's the finished product. What's on this final level?

To answer this question, we will go back to the layout of the human body.

"Voice", "Heart" and "Sex"

We looked at the correspondence between the four elements and the body, but some parts were missing. We had brain, lungs, guts and limbs, but that was all. Between those parts, we find three significant organs, the mouth, heart and genitals.

With a little thought, it becomes apparent that these three body parts represent a mixture of the organs / elements above and below, which leads to a very satisfactory solution.

If we redraw the table with the missing parts, we get the following list with three new entities.

Seven Principles	Description	Body Part
Fire	Will, Desire, Intent, Purpose	Brain + Nervous System
Voice = Fire + Air	Reflection, Comparison, Choice, Logic	Mouth + Vocal Tract
Air	Law, Plan, Information	Lungs
Heart = Air + Water	Alternation, Cycles, Time	Heart + Circulatory System
Water	Work, Action, Movement	Guts, Stomach
Sex = Water + Earth	Mixing, Combination, Sex	Genitals

	Earth	Matter, Object	Limbs

I'm going to call this arrangement the "seven principles" (7P) to give them a distinct name. They correspond to seven stages or "days" of the archetypal process of creation.

These seven "Earth archetypes", occupying the Water and Earth levels, form a duality with the three "Heaven" archetypes" (One, Yang, Yin) which are referenced by the Fire and Air elements.

Understanding "Earth"

This final level of the hierarchy, the Earth level (7), is completely passive. Nothing gets created here because it is the archetype of a finished creation ("God rested on the seventh day").

"Earth" contains (or references) the 4E and the 3Ops, which are created at levels 5 (Water) and 6 (Sex) respectively. It essentially "repeats" them, but in a different format. The whole system is self-referencing, but how should we interpret it when archetypes are repeated by lower levels?

I think we must view each one as a distinct "mind", "perspective", or "view". The Earth element is the "3D material perspective" of reality. It contains (or references) all the preceding archetypes, and it is the only one which does that.

Earth is the (relatively passive) combination of the previous nine (relatively active) principles. It's the final, complete perspective on reality, containing (or referencing / reflecting) all the fundamental, necessary concepts, *i.e.* the whole UP. It's the "finished product".

The Three Operators

In between the existing four elements are places where they "mix" together. This is where the three new principles emerge, and the third dimension is created. These are the "three operators" (3Ops).

Name	Mixture	Between	Organ	Function	Mechanism	Logic
"Voice"	Fire + Air	Brain + Lungs	Mouth	Voice, speech, choice	Reflection	NOT

"Heart"	Air + Water	Lungs + Stomach	Heart	Cycles, time	Alternation	OR
"Sex"	Earth + Water	Stomach + Legs	Genitals	Creation	Mixing	AND

These three new parts are an interface between, and a mixture of, the elements above and below them. They are the mechanisms which can combine or "transmute" the elements from one to another. They strictly define which mechanisms work with which classes of object.

transmute
- to change from one form, nature, substance, or state into another; transform

This "mixing" of the elements is the archetype of (sexual) reproduction and "creation", and the level of the hierarchy this happens at is equivalent to the genitals. It's the sixth level, and it corresponds to the concept of "sex". All three of these archetypes allow a form of "creation".

The 3Ops allow us to create things. They are more "powerful" than their parent elements, with significant new abilities affecting matter. They are more like verbs whereas the 4E are more like nouns. It reminds me of the concept of "hybrid vigour", from biology.

Heterosis, hybrid vigor, or outbreeding enhancement is the improved or increased function of any biological quality in a hybrid offspring.
https://en.wikipedia.org/wiki/Heterosis

The three operators are active principles, like "functions", "powers", or "abilities". They are the archetypes of all possible mechanisms. The UP thus predicts only three fundamental mechanisms exist and can account for all the interactions of matter and minds, from physics to human relationships. It's a bold prediction.

There are only three fundamental mechanisms.

These are "super-concepts" so again, they're still very broad. They include a lot of child-concepts that we might not usually put together, but when we do everything fits well, and all the individual ideas gain more depth of meaning in the process.

The Boolean Logical Operators

The 3Ops correspond to the three basic Boolean logic operators perfectly. This is a stunning correspondence, and it's why I call them the three operators.

Principle	Op.	Name	Description
Voice	NOT	Negation	NOT is REFLECTION. It inverts its input. It takes one argument and returns it's opposite. This is how "one" becomes "two". VOICE is CHOICE: "I want this, NOT that".
Heart	OR	Disjunction	HEART is ALTERNATION. Blood can be in the body-circuit OR the lung-circuit. The two options are mutually exclusive.
Sex	AND	Conjunction	SEX is MIXING / COMBINATION. It's two (or "many") things coming together to make a new thing. The two options are both included.

We can also view them as follows.

Principle	Operator	Mechanism	Description
Voice	NOT Negation, Comparison	Reflection Logic, comparison, division, reflection.	Mechanism of CHOICE Creates the 1st dimension: Properties Allows choice, freewill, decision.
Heart	OR Disjunction, Exclusion	Alternation Alternative paths and places. Cyclic time.	Mechanism of ACTION Creates the 2nd dimension: Forces Allows oscillation, waves, work, power.
Sex	AND Conjunction, Combination	Mixing Mix, join, combine.	Mechanism of CREATION Creates the 3rd dimension: Matter Allows construction, creation, reproduction.

Voice: "Breathing Fire"

Voice is a mixture of Fire, will / desires, and Air, plan *rules* law. It's the archetype of reasoning, choice, decision, and freewill, and of course communication. This archetype, unsurprisingly, has a lot to say. It is our primary route to understanding everything. It expresses the primary mechanism of creation in describable / practical terms.

When we speak, our will is expressed into the air as words. Will is Fire, so this is "breathing fire". When legends say there were "dragons" who could "breath fire" perhaps it meant they could talk. (Perhaps they were people?)

Voice is the mechanism by which choices are made and expressed. We can even use the words interchangeably in phrases such as "giving the people a voice". It is "reflection" and the NOT operator, and it primarily manifests as "negation", *i.e.* "saying no".

Thoughts are a form of language in the mind. There is an internal dialogue even if we're not aware of it. Reason is always expressed in language. We could define the Voice principle as the archetype of "logic expressed", or "reasoning".

This principle is crucial in a universe where individuals have freewill, and that is our perception. How much free will we truly have is debatable, but we perceive ourselves to be free to choose our course of action in life, and the conceptual framework supports that at a fundamental level.

Logic and Causality

Logic is like a thread that joins everything together, it's a special class of concept, one of few that must have existed before duality (in some form). Reason, properly applied, unifies science, religion, art, morality, everything. It is a universal "good". Unreason is the opposite and causes division and confusion, although it does allow for some great entertainment. (Stories always need a villain.)

Freedom for spirit, slavery for matter

Logic could be viewed as providing the duality of "freedom for spirit, and slavery for matter". It allows minds to make choices, giving them freedom, but it "enslaves" matter in "chains" of causality, and "binds" it in the "laws of physics".

While some modern physicists dispute the fundamental nature of causality due to the doctrines of quantum mechanics, duality says it is indeed fundamental and necessary. All things happen by cause and effect.

Causality is the material expression of logic

I suggest that the principle of causality, and the closely related "arrow of time", is a material expression of the logic principle. In other words, causality symbolises logic. Causality connects all physical objects in the universe together by their common causes in a material reflection of how logic connects all ideas.

One cause can have many effects

Causality is a one-to-many relationship, as all dualities are. One cause can have many effects, but an effect can only have one cause. In a game of pool (or snooker), for example, a single strike of the cue can cause many balls on the table to move in complex ways.

The one-to-many nature of causality tells us the universe had a beginning.

If a one-to-many relationship is repeated (iterated) over time it creates a tree shaped structure. Time thus represents a growing logical "tree" of causal events, with a single root cause. Trees have a single trunk because they grow from one seed. One-to-many relationships always originate in "one", unsurprisingly. This must hold true for the universe itself.

The "tree of causality" connects all physical matter, right back to the beginning of the universe. It created and shaped everything we see with our eyes. The "tree of logic" connects all ideas we can see with the mind's eye right back to the "singularity".

All concepts (bar One) also had a beginning.

The UP and its child concepts, just like the physical universe, cannot have existed "forever". It is Yin to One, a creation, which implies it also must have had a beginning.

Everything is made of "circuits".

Logic, as the "law of relationship", connects all concepts together in a hierarchical tree, just like causality does with matter. It connects ideas to their non-physical origin, the concept of "one". One is the source, and the destination, the thing to be explained. Thus, forming a circuit. This is reminiscent of the following:

Revelation 22:13
I am the Alpha and the Omega, the First and the Last, the Beginning and the End.

Duality says everything is made of reciprocal relationships, aka "circuits". This does seem to imply a kind of "electric universe".

Free will is a logical necessity.

Voice is the archetype of free will and we will come back to it at the end of the book. It is a fundamental property of the individual, is logically necessary, and the concept of choice relies on it. We can't just assume it exists; it deserves a full analysis.

The "Christ"?

It seems that the Voice principle has multiple correspondences with the Christian concept of the Christ. It's possible it is the prototype of that religious deity. Generally, the Christ is depicted as a guide or teacher, speaking the truth, but they are firmly linked via the idea of "logos".

We'll explore these connections in some detail as it should help explain the underlying archetype.

In Christianity "Christ" is "the Logos" and is considered to be an aspect of God.

The word "logos" is translated by Christians as "the word", and many seem to imagine it to be a mystical concept, but it has a meaning in philosophy outside of a religious context. "Logos" is usually defined as the rational principle itself, logic / reason.

logos
- (philosophy) reason or the rational principle expressed in words and things, argument, or justification; esp personified as the source of order in the universe
- in Stoicism, the active, material, rational principle of the cosmos; nous. Identified with God, it is the source of all activity and generation and is the power of reason residing in the human soul

nous
- reason and knowledge as opposed to sense perception
- the rational part of the individual human soul

Logic *logos* nous is a true spirit, it has no physical body, but it can be embodied in people. People can symbolise the archetype of reason, by reasoning. Reason must be an intrinsic ability of One as the act of creation requires it. If the "Christ" spirit is indeed the rational principle itself, then that would make sense. It would transform some scriptural texts into statements compatible with science.

John 1:1-4

- In the beginning was the Word, and the Word was with God, and the Word was God. He was in the beginning with God.
- All things were made through Him, and without Him nothing was made that was made.
- In Him was life, and the life was the light of men.

We could rephrase it as follows.

- Logic (must have) existed before the universe was created. It is included in the concept of "one".
- All things were made through logical processes.
- Logical processes created living creatures. Logic is the path to truth and enlightenment.

The above statements are reasonable and not mystical or religious. The primacy of logic and the assumption that the universe can be understood by it is the founding assumption of science.

Ironically, many people seem to imagine the "logos" to be something beyond human comprehension, whereas it is the exact opposite. It's the principle that allows human comprehension.

Is logic a "deity" worthy of "worship"? Perhaps.

We should certainly be glad it exists, without it we'd be "in the dark", unable to understand anything. The "worship" of logic in that sense would be rational. Note, the word "worship" comes from "worthy", and originally meant "worthy of respect" (which is why judges are sometimes called "your worship").

Logic is certainly worthy of respect. We should always consider it and respect its rules.

The "mediator"

The Christ is described by the Catholic Church as a mediator or intermediary between God and creation. Again, substitute in "logic" and it's just a statement of the first assumptions of science, that everything works by rational knowable processes.

- the Logos is an intermediary between God and the world; through it God created the world and governs it; through it also men know God

`https://www.newadvent.org/cathen/09328a.htm`

As the church (and the UP) defines "God" as "truth" (i.e. "reality", "what is"), then we can substitute the concepts, and we obtain a rational and hopefully uncontroversial statement. Logic applied to facts gives us knowledge.

Logic is the mediator / intermediary between the facts of the world and knowledge (truth).

Christ archetype

The Voice principle is probably the archetype of the "Christ" deity / avatar. It could be seen as the "reasonable man speaking truth", the "teacher" or "guide". It creates the Air element, which is law, "a guide". The Christ is considered both a guide and a law-giver.

The spirit of reason can be embodied by human beings. People can be logical and reasonable, which is "Christ like", or they can be illogical and unreasonable, which would therefore be "anti-Christ like". Unreasonable people certainly are "challenging".

This seems to be a plausible, non-mystical interpretation of the concept of the "Christ". A Christ is then someone who is completely rational, reasonable and "sane". Someone who can think freely without falling into the "sin" of ideology, perhaps. The "Christ-mind" would be the "mind of the scientist", interested only in discovering truth.

Taken to its natural conclusion, reason leads to enlightenment.

There is "Christ-like" symbolism in King Arthur receiving Excalibur from the Goddess, or Saint George slaying the dragon. It is the archetypal "triumph over evil", the victory of the Yang over the Yin mind.

Adversary archetype

The "anti-Christ" then, is then one who is irrational. One who cannot be reasoned with, and hence holds illogical, even dangerous, opinions. This could perhaps be characterised as the mind of the ideologue, zealot, or "fanatic". A person who refuses to reason is refusing communication, dividing themselves from the world and refusing to cooperate with it.

The only route to truth is via reason. To refuse it is to refuse truth, to be an "adversary" to it. Christians believe God is "Truth", and "Satan" is "the adversary" to God, also that Satan is the "prince of lies", *i.e.* untruth. So, these ideas correlate quite easily.

Reason / emotion

The fundamental opposite of "logical" is not necessarily "illogical", it depends on context. "Illogic" is more like a "side-effect" of another process. Everything is made by logic, so what seems to be illogic must actually be logic with different premises.

The true (causal) opposite of "logical" is "emotional". Apparent illogic arises from emotion. Intellect and emotion correspond to the Air and Water elements, which are mirror images. Air is logic, it's straight and simple. Water is emotion, and it's circular and complex. People hold illogical views for emotional reasons, aka "feelings". The "ideologue" holds their views because they "feel right", not because they are right.

Yin / Yang minds

This leads us into the topic of the Yin and Yang minds.

The archetypes dictate there are two basic types of mind or "mind-sets": the Yin and Yang minds. The Yang mind is rational, the Yin mind is emotional, and we all experience at least some aspects of both. Everything ultimately works by logic, so emotion has its own version of logic which ends up being the dual opposite of rationality via the "paradox of Yin".

The Yin mind is the selfish, emotional "mind of matter". It corresponds to the "reptilian brain".

The Yang mind is the altruistic, intellectual "mind of God", which corresponds to the "mammalian brain".

Emotions are naturally in conflict with intellect. We know what we should do, but we end up doing what our emotions want instead. People act in the "heat of the moment", only to regret it later. The duality between these two minds is profound.

(Water) Yin Mind – Emotion *Matter* Body	(Air) Yang Mind – Intellect / Spirit
Wants to be "seen as good" (indirect)	Wants to be good (direct)
Pain and pleasure: amorality	Right and wrong: morality

Living in the past, short-sighted	Living in the future, long-sighted
Emotions, feelings, physical urges	Facts, truth, spiritual rules
Selfish: focus is the self, ego, "doing what I want"	Altruistic: focus is others, "doing what is right"
Illogic, circular reasoning, fallacy, paradox	Logic, reason, straight, direct
Competition. Survival when alone The "reptilian brain" / subconscious	Cooperation: Survival in a group / society The "mammalian brain" / conscious
Darkness: absence of knowledge, slavery	Enlightenment, knowledge, freedom

The Cross

The symbol of the cross is associated with Christ. The Church says this is because Jesus died on the cross, and it's a symbol of his sacrifice. However, the symbol of the cross can also be associated with the principle of choice and freewill.

The cross can represent the "crossroads", a choice of which direction to take. In fact, when you reach a crossroads, you are compelled to make a choice: go left, right, or straight on. You are forced to choose, but the choice is yours. It's very Yin/Yang.

You can metaphorically "reach a crossroads" in your life, where you must make a key decision. Also, it's a popular legend that if you want to sell your soul to the devil, you go to a crossroads (somewhere). The cross could thus be interpreted as symbolising freedom of choice.

Etymological dictionaries say there is no link between the words "Christ" and "cross" (or "choice"), but they do link it to "crust" and "grime", so I'm sceptical. This is an investigation for another day though.

The Sword

The cross can also symbolise the sword, and vice versa, as they have the same form. The sword can also represent logic in the sense of dividing truth from fiction; a sword of logic can "cut through lies". A perceptive mind is said to be "sharp".

The sword can also be seen as a symbol of independence, strength and self-reliance, and those properties also epitomise the principle of logic. Logic is independent of nature, it is "stronger" than any other concept, having more "power". It is the pinnacle of self-reliance, standing alone, dependent on nothing yet shaping all things.

Logic leads to knowledge, which gives us freedom and makes us independent and strong.

In the Biblical book of "Revelation" the Christ is said to have a double-edged sword coming from his mouth which "cuts both ways". This symbolises duality, logic, and the Voice principle remarkably well.

He had in His right hand seven stars, out of His mouth went a sharp two-edged sword ... Rev 1:16

When King Arthur retrieved the sword Excalibur from its hole, he was symbolically releasing his power of choice from the clutches of the ego, the "mind of matter".

Also:

- A sword is made of shiny metal and can act as a mirror, and logic begins with "reflection".
- The seven stars in his hand probably represent the seven principles (7P).
- Another obvious interpretation of the cross is the seasons *quadrants of the circle* sinewave.

Heart: Cycles

The Heart principle is a mixture of Air and Water, and the physical heart pumps a mixture of air and water (blood) around the body. The heart is like the body's clock, constantly beating a cyclic rhythm. The Heart-principle contains all concepts relating to time and cycles, alternation, and repetition.

The heart of a thing is its centre / core (Note: French "coeur", Spanish "corazon" etc.)

The Heart principle is thus the "core mechanism" of reality. It brings creation into motion.

heart
- the chambered muscular organ in vertebrates that pumps blood...
- the vital centre and source of one's being. The most important or essential part.

The first active principle, Voice, is like a simple reflection in a mirror creating two co-existing things, the original, and the reflection. The Heart is like an alternating reflection, it's switching between two states which can't co-exist, they're mutually exclusive.

While Voice is a time-less NOT operator, Heart, the OR operator, is like a NOT operator in time or sequence. It gives the choice of "not this" or "not that". "Or" implies a choice between two alternatives, and the physical heart provides two alternative paths for blood to flow, *i.e.* the "lung circuit" or the "body circuit". This symbolises the fundamental principle which is "alternation", or "alternative paths".

The Voice principle creates a complementary pair, the original and the reflection. The Heart creates an opposing pair: high and low pressure. You can have both original and reflection present at the same time, but a liquid can't be both high and low pressure at the same time (relative to the same standard). The two parts of opposing dualities are mutually exclusive.

So, the primary type of duality is complementary and it's created by Voice, opposing dualities are secondary and are created by Heart.

The Heart-principle is of central importance, all the work of the cosmos is done here. If the theory is correct, this is the only conceptual mechanism available for doing work. It needs a more in-depth discussion as it's a bit complicated, so it has its own chapter.

Sex: Creation

I could perhaps have named this principle differently, but it is what it is.

Sex is a mixture of Water and Earth. We could perhaps interpret it as "salty water", *e.g.* the ocean, the "cradle of life" where biological life is thought to have begun.

The Water-element is "motion", Earth is "matter", and Sex is a mixture of the two, so it is "matter in motion". Sex brings material bodies into motion, and it creates new people, who are literally "matter in motion". When you create anything, from a clay bowl to a computer-program, you are making things made of matter move. You are imparting motion / change to a physical thing.

We can describe the Sex principle as "Earth in motion", and I wonder if this is why the phrase "Did the earth move for you?" exists in relation to sex? Perhaps it was an alchemists' in-joke?

In Latin the number six is "sex", the name "sextus" means "sixth". The words sex and six are thus related, and the UP perhaps shows the original reason why, *i.e.* the genitals are the sixth principle when counting down from the head / Fire.

In the act of sex, two things come together and produce a third. It's about combination, taking multiple things which already exist and making something new from them by mixing / combining them.

Sex is a many-to-one operator, a logical AND.

All biological lifeforms have one thing in common, they reproduce or "propagate". The Sex-principle contains all concepts relating to the physical act of creation, not just of babies but anything. All acts of creation "bring Earth into Motion", *i.e.* mix the Earth and Water elements.

propagate
- to transmit (characteristics) from one generation to another
- (biology) biology to reproduce or cause to reproduce; breed
- (physics) to move through, cause to move through, or transmit, esp in the form of a wave

The ability to self-replicate is a defining feature of living things, but this suggests the same principles will apply universally to all acts of creation and reproduction, not just sex between living things. It would also apply to things we don't consider to be alive such as in the propagation of light or sound.

The principle of propagation is to make a copy of yourself, or something else. Gardeners propagate plants. In physics, we can talk about how light or sound propagates, *i.e.* how it travels. Perhaps this is because light and sound does conceptually have to "copy" itself from one location to the next. Sound waves are being duplicated from one "generation" of air-molecules to the next.

Seven Days of Creation

Why are there "seven days of creation" in the Bible, and why do we have a seven-day week? This theory suggests these ideas originated in a knowledge of the seven principles.

If a process of creation has seven steps, then maybe the working week has seven days to reflect that. After all, working is the process of creation. It makes sense it would be designed to be in harmony with the natural structure of the process of creation itself.

A "day" is just a "period" or a "stage". All acts of creation involve multiple stages, forming a sequence. Instead of assuming a specific act of creation, let's just consider the process itself in general terms.

Element	Concept	Question	Stage
Fire	Will	Why	I want a drink
Voice	Choice	Which	I choose to make tea, NOT coffee
Air	Plan	How	1. Put the kettle on, 2. Get a cup, etc...
Heart	Time	When	Now
Water	Work	Who	Me in action
Sex	Creation	Where	In the kitchen, mixing ingredients
Earth	Product	What	The cup of tea

The biblical seven days, and the reason we have a seven-day week can be explained by this arrangement / pattern. The Earth element, "day" 7, is passive, it's "unable to do anything", and the Biblical seventh day of creation was a "rest day". It seems unlikely to be a coincidence.

The Archetype of Genetics?

This final generation of entities also seems to complete the archetype of genetics we discussed in relation to the four elements. The way it works doesn't look quite the same as physical DNA, but it does have similarities.

The three-operators still have two "chromosomes" X/Y, but the genes are more complex than the previous generation. The 4E's genes were either Yin or Yang. In this new generation the genes are the four elements. So, now we have four possible options instead of just two.

This is the completed archetype; it has these features. - A pair of "chromosomes" containing a single "gene" each - The gene is made of a single "base-pair" of two "nucleobases". - There are four possible different "nucleobases" (FAWE vs ACGT in DNA).

	X Chromosome	Y Chromosome
Voice = Air + Fire	Air (+-)	Fire (++)
Heart = Air + Water	Air (+-)	Water (-+)
Sex = Earth + Water	Earth (--)	Water (-+)

Note the genders remain consistent between pairs and within the framework, Air is X/Feminine, Water is Y/Masculine, and these qualities align with the active/passive categorisation of the four elements we will see when looking at the Heart process.

This could be the simplest-conceivable fully formed version of a type of genetic inheritance. It's similar to physical DNA, but there are differences. The physical version is a lot more complicated, and its "indirect": it codes for proteins which have properties, rather than coding directly for properties as we see here.

This opens the way for a family tree of concepts to develop from this core template. The "spirit world" is populated by families of concepts / minds who inherit their properties from their parents using an inheritance mechanism similar to that employed by physical creatures.

"As above so below", I suppose.

Parts of Mind

We have described the different parts of mind as "internal senses". Now we have the complete plan including the 3Ops, so how do the various parts of mind fit into it?

Note:
- Physical senses are in bold.
- This table is a work in progress. There's room for improvement.

Category	Physical Sense	Spiritual Sense / Mind
One		Sense of identity. "Me" Being, consciousness, existence, awareness, self
Two		Sense of "correctness", conscience, intuition... The Yin and Yang minds. Yin: Instinct + survival when alone. Yang: Learning + survival in a group. Conscience / morality
Fire - Will	**Vision**	Sense of Desire
Voice - Choice		Sense of Reason(ing) Reflection, comparison, imagination, thinking, choice Logic, information-processing, language, communication
Air - Law	**Hearing**	Sense of Knowledge Understanding, judgement, thoughts, rules, ideology, beliefs, plans True / false. Right / wrong
Heart - Time		Sense of Time Future / past, timing, urgency, regret, hopes + fears, duty
Water - Work	**Smell / Taste**	Sense of Emotion Emotions, feelings, ego, happiness, sadness, fear, jealousy, grief, *etc.*
Sex - Creation		Sense of "Like". Sense of Humour Sexual/physical desire *like* attraction, (also disgust...?) (Sex is bringing two things together, and so is humour)

| Earth - Matter | Touch / pain *etc.* Hunger + thirst, health... | Sense of Memory Memory, recognition, recall, facts and data. |

The three operators' give us mixtures of the physical and spiritual senses. This is perhaps most apparent in sexual desire which is observably a combined mental and physical sense. The 4E senses are all very much "divided" senses, whereas the 3Ops bridge the gap between the inner and outer worlds.

Voice is an interface between our inner senses of self, desire, and knowledge. It deals with things in the world that we need to judge / compare.

Heart interfaces between our knowledge and our work / duties. It connects to time in the real world, giving us our sense of "urgency", among other things.

Sex interfaces between emotions and the physical world. Its spirit is about joining things together because it feels good and we "like" it. It is the archetype of all kinds of creation, from making babies to writing symphonies. It includes making jokes, making food, and all kinds of invention and novelty.

It seems "sense" is one of the most complex ideas that exist, spanning all the archetypes.

In total there are thirteen different classes of sense, but some of those classes contain multiple individual senses. A full analysis would require a more detailed view of each category and would take the form of a nested list, but that's beyond the present scope.

Conclusion

We have now covered all ten of the basic archetypes within the system and all the core theory.

The whole plan was derived from first principles, albeit with a bit of induction to help guide the way. We have ended up with a system that seems to be complete, logically consistent, descriptive, and predictive.

The UP appears to be logically necessary, arising directly from the most basic rules. It's an inevitable result of the logic principle and describes its underlying form. The UP is like logic's "body". It is the "shape" of logic, information, language, and everything else. We could say it is the full definition of the "logos".

The next stage is to apply the rules we've discovered to some more real-life phenomena, to see if the theory is any good in practise.

Note: The "Gods of Nature"

The UP's template of archetypes could explain the origin of some traditional belief systems.

The idea that there are "gods" both outside and within nature is a common feature of polytheistic religions, and the UP allows us to "rephrase" or "reframe" the concept in relatively scientific terms.

The ten entities the UP describes can be considered as "gods" or "deities". One, Yang and Yin are the "gods in Heaven", and the 7P are the "gods of Nature" (or "Earth").

The UP is a list of "deities". Every archetype in the UP can be considered as a "god", a "divine principle", and I suggest they are the original prototypes for at least some of the pantheons described in ancient cultures around the world. For example, "Odin" means "one" in Russian and Sanskrit (aka "Woden" in Britain).

The nine lower principles can be thought of as One's "children". The UP is "God's family tree".

The seven Earth principles are presumably also the "seven spirits of God" mentioned in the Bible.

```
https://en.wikipedia.org/wiki/Seven_Spirits_of_God
```

Non-mystical deities

It's important to remember that these archetypes are essentially just "information". They are a form of data, or at least can be described by it. There is nothing mystical, nothing beyond human comprehension or experience here.

The "gods of nature" are the fundamental archetypes reality is constructed from. They are the foundation of the world of concepts. All the principles in the UP equate to completely rational, explainable, listable sets of properties which anyone can understand.

Although, having said that, they must also be living minds, each with its own perspective on reality. But minds are not mysterious, we all have one.

Note, the official etymology of the word "day" specifically says it is not linked to the word "deity", but I think this is a mistake. The day is conceptually a "provider", a "god". We find that the word "deity" originates in the verb "to shine", which obviously links to daytime, as opposed to night.

deity: from PIE deiwos "god," from root dyeu- "to shine,"

`https://www.etymonline.com/search?q=deity`

Fundamentally, the daytime symbolizes Yang *God, and the night, Yin* Goddess.

Chapter 6

Language

The UP describes the fundamental structure of all things, and in a non-physical universe that really means "information". It is the "master plan" of the structure of logic, information, and thus "language".

A theory of everything would have to be something like a "language of the gods" in some way and the UP is overtly such a thing, it's a language "made of gods". If this theory is correct, it should be the underlying structure of all language.

So, how well does human language match the Universal Plan? If it provides a reasonable match, that would be good evidence in favour of the theory.

In summary, it matches well, and I suggest provides a much better understanding of the phenomenon than the current paradigm can offer. Not only can we better classify language, we can even unite human and computer languages, which I believe is a first. This seems to be strong evidence demonstrating the utility of UP theory.

It took a few iterations to figure out how this could all work coherently. I will spare you the details of the process, but I should point out that some of this wasn't immediately obvious, and it is still open to improvement.

Language contains some dualities which are worth a mention but classifying them is beyond the present scope. There is more structure in language than we will be able to get into here.

- Open and closed classes of words:

Closed word classes only contain a limited number of words, *e.g.* conjunctions. Open word classes contain an unlimited number of words, *e.g.* nouns.

- Function and content words.

Content words contain meaning and belong to open classes, *i.e.* nouns and verbs.
Functions words join and relate content words, and belong to closed classes.

- Two types of question.

There are questions that require a yes/no answer: *e.g.* "Is that true?"

There are questions which invite a free response. The "wh-" questions: Why, where, *etc.*

The Parts of Speech

Different types of word are called the "parts of speech" or "word classes".

There isn't universal agreement in academia on what the fundamental word classes should be. Wikipedia lists eight or nine, but there is some overlap between the classes, and they aren't necessarily supposed to all be distinct.

- noun
- verb
- adjective
- adverb
- pronoun
- preposition
- conjunction
- interjection
- determiner

https://en.wikipedia.org/wiki/Part_of_speech

How might these different categories correspond to the UP? We have ten possible slots in the hierarchy to fill, but only nine word classes. One of those will be disqualified, and two will be combined, leaving seven. However, it turns out it's not as simple as just allocating them to the 7P.

Parts of Computer Language

To crack a problem, it can be useful to draw parallels with similar systems, and computer language could be just the "missing link" we need. It's an extra data set, another example of the phenomenon of language that could be used to provide more context. Human language is complex, it's evolved over time, and even the experts disagree about its structure.

The UP says computer languages (CL) must use the same basic template as human languages (HL). CL should be simpler than HL as they only explain a subset of the things the latter does, and they are deliberately designed to be efficient and understandable.

I'm going to use the correlation between these two language-classes to help explain both better, to attempt to identify the "spirit" of each of the categories, so we can define them in simple terms. I couldn't find any prior art correlating the "parts of speech" in human and computer languages.

HL contains the fundamental duality of noun / verb, which I've used a lot in this book. Verb is Yang because it's active, and noun is Yin because it refers to objects.

The first question we must answer is: what is the most fundamental duality in programming that could correlate to nouns and verbs? The answer is simple: data and instructions.

	Yin	Yang
Human Language	Nouns	Verbs
Computer Language	Data	Instructions

Computers process data with instructions. That's the essence of what they do, it's their core duality, and it easily corresponds to nouns and verbs in HL, and directly to the Yin and Yang categories.

That's the first piece of the puzzle solved, and two categories filled. The next step is to determine what are the most basic word-classes are in CL.

Wikipedia provides a list of the different types of "statement" available in CL here:

```
https://en.wikipedia.org/wiki/Statement_(computer_science)
```

If we "abstract the essence" from the list of "simple statements" provided, it fits into the 4E very well. Although the correspondence isn't as obvious as in some other cases, it's still there.

The most basic "parts of speech" in computer languages are as follows.

Element	Type of operation
F: Will	**Flow control: loops, branching, direction** if/else, for/each, while, do/until, break, end... These determine which part of the code gets executed. They "direct" it.
A: Law	**Logic: comparison, relationships** Always return a TRUE/FALSE. (e.g. ($x>$y), ($book.Due==TRUE) *etc.*

		Is always a "test of properties", a measurement, a comparison against a standard.
W: Action	**Functions: data manipulation** May/may not return any kind of data $x=function(\$x, \$y)$ ... Functions are where all the work is done. There is an "ocean" of possible functions.	
E: Matter	**Assignment: data, variables, constants, values** Data-handling is the goal of all software. Data / information is the "matter", the result, the end-goal, the "value".	

This is another part of the jigsaw that looks completed. The next thing to do is correlate human language to the list above. Then we can compare the two and see if they match.

If we subtract nouns and verbs from the list of word-classes, we need to account for the remaining:

- adjective, adverb, pronoun, preposition, conjunction, interjection, determiner

Interjection and Informal Language

"Interjection" is sometimes listed as a separate part of speech, but I will argue it's more like a different type or "mode" of language.

Interjection ... is a diverse category, encompassing many different parts of speech, such as exclamations (ouch!, wow!), curses (damn!), greetings (hey, bye), response particles (okay, oh!, m-hm, huh?), hesitation markers (uh, er, um), and other words (stop, cool).

`https://en.wikipedia.org/wiki/Interjection`

Chatting about football with a friend is a very different type of communication than explaining a theory for a scientific journal or defending oneself in court. We use different "modes" of language for different situations.

Humans have two basic modes of language and we could describe them with the dualities "formal *informal*" or "*intellectual* emotional". They are like two different languages, designed for different purposes. Informal language is appropriate when communicating emotions. Formal language is better when communicating facts.

Formal language is designed to be unemotional, to convey its ideas via intellect. It's good for communicating long detailed facts, remotely, indirectly, in written form, and to people you don't know. Informal language is designed to express emotional content, feelings, and humour quickly. It doesn't convey as well in written form.

Formal: unemotional, intellectual, dry, facts and figures. Long and detailed. Informal: emotional, heartfelt, feelings. Short and abrupt.

Interjections are a subset of the informal mode of communication. Expressions like "wow" or "no way" can be expressed in formal language, but using the informal mode is more "friendly".

Anyway, this means we must disqualify interjection from the list because it's not a class of word but (part of) a different language "mode".

Determiners and Pronouns

After some deliberation I'm going to suggest that determiners and pronouns are two side of the same coin and belong in the same category, as a duality. They are both essentially nouns which refer to nouns, but in different ways, from different directions.

We have to look at the problem functionally: what problem are these word classes attempting to solve? What is their purpose?

I suggest it's this: in both human and computer languages we often need to be able to select a set of data (i.e. specify a group of things), and then refer to it as an entity. With determiners and pronouns, we can do just that; we can collect an arbitrary set of data and refer to it as a unit.

"Selectors"

Determiners combine with other word classes (e.g. prepositions and adjectives) to create what we might call "selectors" which specify a set of data. For example, "all the brown cars in my lot" or "your youngest Australian cousin". The simplest selector is "the", which is strictly a "determiner", but they can include other word-classes.

selector

- a phrase which identifies a set of data, which can then be referred to via pronouns

"Selector" statements are functionally equivalent to retrieving data from a database with a SQL "SELECT" statement. Their purpose is to identify a set of data according to a set of rules. They can "return" one or more "rows of data", *i.e.* refer to one or many individual items.

Pronouns are like the variable the data is stored in, *e.g.* $they = {"All the brown cars in my lot"}. They allow us to refer to the dataset conveniently. The determiner / pronoun relationship is thus equivalent to the relationship between a variable and the data it refers to (is "assigned").

Human	Computer
Determiner / selector "All the brown cars in my lot"	Query $cars=Get("SELECT * FROM cars WHERE colour='brown'");
I've decided to sell them.	MarkForSale($cars);
But they need a good wash first.	ScheduleValet($cars);

Selector *pronoun is a complementary duality equivalent to value* variable assignment. Selectors come first, providing the definition. Pronouns come afterwards, referring to it. There can be one selector to many pronouns. The pronoun/variable is passive, the selector is active, *etc.* thus selectors are Yang to pronouns.

In pseudo-code we might write the relationship as:

`$pronoun = GetData($selector);`

Determiners / selectors are complicated because of the difficult function they must perform. Choosing a set of data from a universe of possibilities can involve a long list of descriptions and rules. It seems likely that this function in human language is directly analogous to an SQL SELECT, and the two systems can and should be correlated.

So, determiners take the Earth slot in the 4E. They correspond to assignments in CL, and we've already established they belong under that category.

Note, because determiners and pronouns occupy the Earth category, we might expect selectors to reference all other word classes, as the Earth archetype references all the nine higher ones (covered in "Understanding Earth" in the previous chapter). As this is the most Yin category, it is also the most complex.

To have both concepts of "determiner" and "selector" is a bit messy, one is probably redundant. The ideal solution would be for "determiner" to have all the abilities the proposed "selector" does. So, we would need to allow for this in the definition of "determiner" somehow.

A possible approach might be to treat "determiners" not as a word class, but as a structure which can contain or "invoke" all other classes, I'm not sure. How we might integrate these ideas is a question for future consideration, although it's probably not too difficult. It may be solved by an analysis of the SQL SELECT statement.

Adjective, Adverb, and Questions

We've dealt with five classes out of nine: nouns, verbs, interjection, determiners, and pronouns.

The outstanding list now consists of four entities: adjective, adverb, preposition, conjunction.

Preposition is "location", and so I have that earmarked for the Sex principle (Earth + motion). Conjunction is logic, so that would fit under Voice. That leaves only adjective and adverb, but we have three slots to fill in the 4E, and also Heart in the 3Op is still unassigned. There must be another two classes, that aren't currently recognised as such, to discover.

Firstly, lets focus on getting the 4E assigned.

The four elements are made of a defined combination of Yin and Yang. The four word-classes we're looking for will fit that pattern, we just have to determine what the fundamental qualities of the concepts we're dealing with are, and "crunch the numbers".

The basic properties of nouns (Yin) and verbs (Yang) are "form and function". The 4E inherit these properties from them and alternate them. Form / function is a complementary duality, so we can plug it straight into the algorithm. I'm interpreting Yin *Yang as defined* undefined, as it seems to be the clearest way.

	Yin: Noun, Form -	Yang: Verb, Function +	Word Class
F ++	+ Form is undefined	+ Function is undefined	Questions (Interrogatives) Words calling for a definition. Everything is undefined. A "verb-verb"
A +-	+ Form is undefined	- Function is defined	Adverbs Adverbs describe verbs. Function is defined, but not form. A "verb-noun"

W -+	- Form is defined	+ Function is undefined	Adjectives Describing physical things / appearance *etc.* Function is not defined, but form is. A "noun-verb"
E --	- Form is defined	- Function is defined	Determiners + Pronouns Nouns identifying other nouns. Both form and function are defined. A "noun-noun"

The first missing word-class is questions, "interrogatives", and it is the "Fire" category. Question words are not currently considered an independent class of words, but this says they should be.

To check the correspondence is right we can now compare the two lists for HL and CL.

Under "Fire" for computer language is "flow control", and that is the same concept as asking questions. Flow control is about finding the answer to a question, so you can act on that information. So, this is a match. *E.g.*

```
// Is $x greater than ten? If so, end the program.
if($x>10) goto end;
```

Adverb also fits perfectly here. Adverbs describe an action, which is the same as limiting an action, which is the same as "law", which is Air. Laws limit action. In the sentence "he ran fast.", fast is the "rule".

Adverbs "modify verbs" and adjectives "modify nouns" by making them more specific and detailed, and thus more "limited".

A limit on action is a "law". A limit on matter is a "property".

Adjectives and adverbs are clearly mirror images of each other, as both are "describing" classes. Questions and determiners are also a reflection of each other. Determiners / pronouns answer questions. A question like "is x greater than ten?" is answered by "x = 12".

3D Human vs 2D Computer Language

We can now correlate the two types of language, at least to the level of the four elements. This shows some differences between HL and CL, with HL having greater flexibility in being able to modify verbs.

	Human Language	Computer Language
F	Questions if, why, which, how...	Flow control if-then, while, for-next...
A	Modify / compare verbs Adverb: verb descriptor, "rules". - fast, faster, fastest...	Compare data True/False answers. ($x > $y), ($x < 100)...
W	Modify / compare nouns Adjectives: noun descriptor, "properties". - red, redder, reddest...	Modify data Functions Changing data. Work.
E	Determiners + Pronouns	Values + Variables

It's interesting how different some ideas are between HL and CL, and yet they do still fit together. The amount of new information revealed by the correspondence is considerable because of that wide difference. I think we can learn a lot about both forms of language, and the 4E, via this comparison.

One difference that emerges is, it's not possible to modify computer instructions or mathematical operations as you can verbs. A human can run "fast, lazily, easily, or painfully" but computer instructions lack this ability to be modified. A computer can multiply two numbers, but it can only do it one way. There is no conceivable modifier for a basic mathematical operation like plus, minus, divide and multiply, or for Boolean logic operations.

In other words, human language has an extra dimension, an extra degree of freedom. Presumably HL is designed for communicating 3D things, and CL is designed to communicate 2D things. Quantitative computers are two-dimensional devices.

The final step to complete the analysis is to add in the three operators and figure out how they are represented.

Time and Tenses

There must be three classes of word in HL which correspond to the three operators. They must also correspond to some analogous features of CL we haven't yet identified.

Voice is "conjunctions" because that's where all the logic operators like AND / OR go, and Sex is "prepositions" which are locations.

Heart is time, and it took a moment to figure out what class of words should go here because they aren't recognised as a fundamental word class. The Heart principle, as the archetype of time, must be something to do with verb tenses, how actions are positioned in time. It must be a class of words which specify the tense of a verb.

In English and many other languages, the tense can be conveyed as part of the verb, especially for common verbs and tenses (e.g. "I wrote"), but generally we use what are called "auxiliary verbs" to provide the time-context, along with some suffixes on the main verb.

`https://en.wikipedia.org/wiki/Auxiliary_verb`

For example, the verb "to work", with 12 tense-options.

Tenses	Past	Present	Future
Simple	I worked (yesterday) (I did work)	I work (most days)	I will work (tomorrow)
Perfect	I (had) worked	I have worked	I will have worked
Continuous	I was working	I am working	I will be working
Perfect Continuous	I had been working	I have been working	I will have been working

The take-away here is that the time content could be conveyed exclusively by "auxiliary verbs". The body of the verb could remain unchanged throughout the tenses, while only the auxiliary changes. This would better fit the archetypes of the UP, so it suggests it might be the more "natural" way to specify tense, as opposed to changing the verb itself.

The three operators allocate as follows, and this almost completes the picture.

Voice - Logic	Logic-joiners: conjugations
Heart - Time	Time-joiners: auxiliary-verbs
Sex - Location	Location-joiners: prepositions

The only remaining question is what parts of CL do the 3Ops correspond to. Those gaps are filled in below.

A Map of Language

This is the final map of concepts *word-classes, and it includes all ten entities in the UP. It shows the four basic levels, with One at the top and the three operators on the final level* generation.

The system begins with the verb "to be". "One exists", and that's the only language available at that point. The first statement is "I am", where "I" is a noun and "am" is a verb. "I exist" is the first, and arguably only, fact that any consciousness can know for certain.

This concept gets divided into Yin and Yang, noun and verb, matter and spirit.

The four elements define the next level, which are more complex concepts that help describe verbs and nouns in detail. The final level defines how connections between concepts can be made. This is the table shown in the Overview in the Introduction.

Level 1. One, unity HL: "To be" / "I am" CL: The executive: "run", "exec"...			
Yin - Matter, solid objects HL: Noun CL: Data		Yang + Spirit, abstract objects HL: Verb CL: Instruction	
Earth -- HL: Determiners CL: Assignments	Water -+ HL: Adjectives CL: Functions	Air +- HL: Adverbs CL: Comparison	Fire ++ HL: Questions CL: Flow control
Sex HL: Prepositions CL: Math operators	Heart HL: Time-Joiners CL: Code-blocks	Voice HL: Conjugations CL: Logic operators	

The table shows the 7P part of this system in more detail with some examples. We can also correlate the individual question words with the 7P, and they fit very well, adding excellent context.

	Human Language	Computer Language	Example	Question
F	Questions why, which, how...	Flow control if-then, while, for-next, goto...	if	Why

V	Conjunctions and, or, but, not, nor...	Logic operators and, or, not, <, >, ==...	>	Which
A	Adverbs Verb modifier, "rules" - fast, faster, fastest...	Comparison if($x > $y), while($x < 100) ... Modify actions	$x > 10	How
H	"Auxiliary verbs" Time / tense modifiers Alternative times	Code-blocks, "compound-statements" Alternative code paths	if ($x > 10) then code path 1 else code path 2	When

W	Adjectives
	- noun modifier, "properties"
	- red, redder, reddest...
		Functions
		Modify data
			function update_x ($x) {
			$x = $x + 10;
			}
				Who

X	Prepositions, locations "Earth / nouns in motion"	Maths, strings... Mix data, create new data	$x=$x + 10;	Where
E	Determiners and Pronouns Noun selector / specifier	Values and Variables Data specifier. Assignment.	$x = 10;	What

Notes:
- "Not" is not considered to be a conjunction, but it should be.

Three operators in CL

The 3Ops correlate to CL as follows:
- Voice: Logic operators correspond to Voice, as expected.
- Heart: Compound statements fit in the Heart category because they represent alternation, *i.e.* alternative paths for the process to take.
- Sex: Mathematical operations fit under Sex, they allow arbitrary "mixing" of

numbers. It's interesting maths would appear under this category, it's not a bad fit. Food for thought.

The hierarchy begins with "I am" at the top and ends up with "I am equal to ...".

It all seems to make sense and ends rather neatly. The purpose of the universe is to "explain what One is equal to", and that's exactly where we end up. The plan fits well, includes all the parts of language, and provides links between related concepts that would otherwise have been overlooked.

There is clear linkage between the basic principles, the parts of speech, and the question words. They seem to belong together; to not have these links made explicit would diminish the meaning of all the concepts involved. This is, I believe, more good evidence in favour of the theory.

These last two sections aren't necessary to make any point but are included for discussion.

Types of sentence

There are four types of sentence structure. Looking at their most basic properties, they seem to fit into the 4E quite well, although the assignment of the two middle elements is debatable.

Element	Sentence Type	Description
Fire	Simple	One clause
Air	Compound	Two independent clauses (Like two separate molecules of gas?)
Water	Complex	One independent, One dependent clauses (Like attraction between molecules in a liquid?)
Earth	Compound-Complex	Mixture of above

Ordering Adjectives for Nouns

There is a natural ordering for types of adjectives, for example we would say "the big red metal bridge", and not "the metal red big bridge". There does seem to be correspondence between them and the UP, but it's not perfect.

https://dictionary.cambridge.org/grammar/british-grammar/adjectives-order

The Cambridge Dictionary lists ten different categories of adjective in the sequence which are shown in the table below, numbered as they appear in that list. They fit in well enough to indicate there is some link between the two lists, but it's not a one-to-one match. Some categories seem anomalous, like "10. Purpose" under Earth.

Perhaps some of this is convention. It's not necessary for all parts of the English language to match the system perfectly. Human languages are, at least in part, the invention of people. Natural phenomena should always match the system, human culture and conventions may not.

Element	Related to	Examples
F - Why	1. Opinion	Lovely, beautiful, intelligent
V - Which	2. Size 3. Physical Quality	Big/small, tall/short, huge/tiny Fat/thin, smooth/rough, tidy/messy
A - How	4. Shape	Round, square, rectangular
H - When	5. Age	Young, old, youthful
W - Who	6. Colour	Blue, orange, pink
X - Where	7. Origin	Japanese, French, southern
E - What	8. Material 9. Type 10. Purpose	Metal, plastic, wood, brick Three-legged, general-purpose, U-shaped Writing, cleaning, cooking, heating

Together with determiners, this sequence of adjectives forms the "selector" phrase which identifies a noun or set thereof, as discussed earlier.

- *"The lovely big square old orange Japanese wooden three-legged writing desk."*

That covers the basic analysis of language. There are still questions to answer and holes to fill in, but hopefully this framework can provide something solid to build on.

Chapter 7

The Heart

This section is somewhat "technical", dealing with various details of the Heart / work process.

The Heart is the gateway between the simple "Heaven" elements of Fire and Air, and the complex "Earth" elements of Water and Earth. It's where things start to get a bit complicated.

The Heart enables:
- The 2D level of reality and the four elements.
- All forms of work, computation / decision making.
- The "third person" perspective and orthogonality, the "independent observer".
- The "organising principle", which is theorised to be the origin of matter and biological life

The Heart mechanism is the "core" of the universe. It's at the centre of everything, sitting between Air/path and Water/action. It is therefore the archetype that lets us "walk the path" and it corresponds to "alternation", just like walking with two legs.

Forgetting God

A fundamental inversion of perspective occurs at the Heart. Yang/God is "inside", but we perceive God as being "outside". We perceive "Nature" (reality / the universe) as "God", it is our provider, and it is external. This shift in perspective is profound, and something of a "miracle", although it's really just another form of reflection.

This is the mechanism that simultaneously "hides God" and creates us (independent observers), allowing us to question the most basic of his attributes, existence. People are free to explore the concept of the non-existence of One. That's quite a feat considering there is only One, and everything is made of it.

Whether someone is an atheist or not, we all necessarily perceive reality as external and provided to us by mechanisms outside our control. Nature is our "great provider", it is tantamount to a "god" even if that label is avoided.

Things that happen to us which are outside our control we see as "acts of God". If a tree falls on a man walking in the woods and kills him, or a hurricane destroys his house, it certainly wasn't his choice, so we reason it must have been "God's choice".

This perception of an "external God", and our independent existence, is enabled by the Heart principle.

Alternation, not "work"

The Heart is the mechanism that enables work, but it's not the work itself. I found it was easy to get them mixed up.

The physical heart does work, and we'll examine it and see what sequence of the four elements it's beating out. First though, a few more definitions.

Mechanisms and Machines

I've been using these terms without a clear definition and can now do so with reference to the UP.

We can use the word "device" as general term for this class of object.

device
- an object that has been devised, invented, contrived, to fulfil a particular purpose
- from Old French deviser "arrange, plan, contrive," literally "dispose in portions," from Latin dividere "to divide"

There are a few closely related concepts which need categorisation so we can define them more clearly. These are all "devices" we use to manipulate matter.

tool
- a device that aids in accomplishing a task

machine
- a device designed do a particular type of work

mechanism
- a process, technique, or system for achieving a result
- a method for converting one force (or element) into another (my definition)

Tool and machine are conceptually distinct. A tool helps you do work, a machine does the work for you. A tool must be guided by your will, it's an extension of it. A machine can be left to do its own thing. I suggest these concepts can be categorised into the 4E as follows.

Note, the way we use these words doesn't have to correspond strictly to this scheme. These are the basic archetypes that (I think) the UP dictates but there can be flexibility in how we use them.

F Will / Why	**Tool** Device with a single purpose. An instrument. May have no moving parts. *Will is a tool*	Sticks and stones Blades and hammers...
A Law / How	**Mechanism** Device to transform / divide forces / elements. Must include distance / separation of in/out forces. *Law is a mechanism*	Lever, wheel, bearing, ratchet...
W Work / Who	**Machine** Device to do work. A collection of mechanisms. Includes motion and rotation. *Work is a machine*	Car, watch, generator...
E What	**Frame / Foundation** Device that provides a reference point / foundation. *Matter is a framework*	Machine casing, fulcrum Chair, workbench, road, walls...

A machine can be a tool, but the concepts have a different focus. "Tool" implies a single purpose, whereas "machine" implies motion. "Mechanism" implies method, "how", and "Frame" is "what".

Note, if a machine is an arrangement of parts designed to perform a task, then by this definition a mind is also a type of machine.

The "Frame"

A "frame" is a kind of device which the 4E demand must also be considered. It was the empty slot of "Earth" which prompted its inclusion, and it makes sense to do so.

The Earth element is "frame" because it is the foundation; the framework of the machine, the fulcrum the lever acts against, or the support the axle rotates within. The frame is an essential part of any device, and a frame can be a device, like a chair, or a set of stairs. It is passive and solid, with no moving parts. It's the "frame of reference", and the fixed origin point around which other things move.

The Earth corresponds to determiners in HL and as variables in CL which are also a "frame of reference", a "frame of mind", a dataset. When we explain things, we "frame" them, and an explanation can be a "framework", such as the UP.

Rotation

Machines are the four elements, and the 4E are a machine. That's why it appears under Water.

Machines are an alternation of Yin and Yang. They have parts in motion around parts which are fixed. The fixed parts provide the solid reference point for the moving parts to move around, so they can transmit and transform the energy / information that is their input into whatever is the output. This says all conceivable machines must include some form of rotation / alternation about a fixed point.

If we consider the basic principle of the lever, we find it fits into the UP very well. This template potentially describes a "universal mechanism", we might find all three of the basic mechanisms / operators fit this pattern.

The Basic Form of "Mechanism"

F	Force, pressure, input, driving will	The top two elements are "Heaven"
V	Reflection: force is divided into two: Input and output.	
A	**The "Stick" / "Lever"** Distance. The force is transmitted through a distance / difference.	Linear, straight, direct

H	Alternation: the force can go in another / alternative direction	*Orthogonality*
W	Motion: rotation about a fulcrum / axis	Curved, indirect, rotational
X	Mixing: bearings / couplings. Join moving and fixed parts.	
E	**The "Stone" / "Fulcrum"** The fulcrum *axis* frame.	*The lower two are "Earth"*

This describes a (relatively) linear force being transmitted across a distance, then converted into a rotational force by moving in relation to a fixed point. The top two elements are linear, the lower ones are rotational. This seems to be the basic template for "mechanism".

(Note, "rotation" includes alternation between two states, such as in digital electronics.)

Of course, we would need to consider all different types of mechanism to see if they really can fit into this archetype. So, this is just a suggestion at this stage.

The Heart Mechanism

The Heart-principle is the centre of the framework; it's the universal mechanism of alternation. It's the origin of the four elements, and all oscillation including light. The Heart is the clock. It counts time and is the origin of Yin-time.

Ultimately, there is a deep link here between the concepts of "alternation" and "pressure", as the mechanism appears as both phenomena. This region of conceptual space needs more exploration than we can cover here.

We must be careful to distinguish between Heart and Water. The work itself is Water, the mechanism is Heart. The Heart is a mechanism that causes waves but it's not the waves themselves.

Why does the output of the Heart process have to be a sinewave / circle?

If we consider the most basic properties of the 4E, the concept of the wave / circle seems unavoidable and occurs under the Water element. Air is 1D, linear motion. Water is 2D and circular, they are mirror images.

The sequence looks as follows.

Principle	Concept
Fire	One, undivided potential
Voice	To Divide
Air	Two, polarity, opposition, 1D - LINEAR
Heart	To Alternate / logical OR
Water	Alternation between two states: 2D The sinewave, circles - ROTATIONAL
Sex	To Combine
Earth	Product, matter.

The Heart must explain how the Water element is made from Air, *i.e.* how to make the four elements from duality, and we have already covered that. It takes a complementary duality, and alternates the properties: (++, +-, -+, --)

The Pump

To figure out how the Heart-principle works, the obvious method is to consider what a physical heart does. It should somehow embody the principle or mechanism we're looking for. If the "heart principle" is well named, it should really be identical with what a human heart does. So, what does a heart do?

The heart is basically a pump. The dictionary says:

pump
- a machine or device for raising, compressing, or transferring fluids

We need a more generalised description, so we can include solid matter and information too. An extreme abstraction is required, such as:

- a device for moving stuff from one place to another

It is conceivable that all types of work might fit under that last definition. All types of work do involve basically just moving stuff around, be it shovelling cement or betting on the stock market. Just as the heart's purpose is moving oxygen from the atmosphere to your cells, all work can be thought of as moving a substance from one place to another. Even intellectual, non-physical work can be described this way.

Pumping out thoughts

To write a book or a program, you have to "pump" immaterial ideas out of your head into the physical world. You're moving the substance of "thoughts" from inside your mind to the computer's data-store (or paper). Every time you think what you want to say, then write it down is like the stroke of the piston. It's a repetitive process of moving stuff from one place to another, (hopefully) increasing its state of order / separation.

Organising a morass of thoughts into a structured linear progression is like stacking bricks on top of each other, or pumping water up a hill. It's increasing the "order" and the "energetic state" of those thoughts.

Before writing, the ideas only orbit around in your own head, a low-energy orbit with low-potential for causing change. Once the book (or program) is written and other people can use it, those thoughts can potentially orbit around the whole world and have a high capacity for causing change.

Entropy, Energy, and Information

There are important links to concepts like entropy and energy here which we will need to explore. The essential problem all physical entities face is energy loss, and they are all energy management machines. This is covered in depth in the section "Light, Energy, and Matter" in the Discussion chapter.

What is notable here is that the Heart principle of orthogonality is the origin principle of physical matter and biological life. The Heart enables the existence of life, and it explains how matter works.

Collision *Impact* Pressure

The definitions of "pump" above describe a result but not a mechanism. They say what it does but not how it does it. All pumps operate by the same principle, and it does create waves. What causes waves in nature but isn't itself a wave?

The most obvious example is a stone dropped in a pond. This mechanism isn't a wave, it's a collision, an "impact", but it creates waves. Plucking a guitar string is also a collision which creates waves.

collision
- a violent impact of moving objects; crash
- a condition of opposition or conflict between two or more people or things

- (physics) the meeting of particles or of bodies in which each exerts a force upon the other

impact
- the force transmitted by a collision
- influence, effect, touch, shock, percussion, impulse

The physical heart only has one mechanism: it can contract, causing collision between the blood and heart tissue. This causes the blood pressure to increase so it flows around the system. (Note: orthogonal to the direction of pressure.) All waves originate in "collisions" or "pressure changes", and they propagate the same way.

When waves propagate through a physical medium they do so by impact. Sound is caused by pressure waves in air, propagated by air molecules "hitting" each other. The same applies with water waves. All work is done via collision *impact* pressure, and waves / oscillations are the result.

A Single Force

If there's only one possible mechanism for work, then there's only one fundamental force in the universe, and it is a "push-force". This would mean all apparent pulling-forces such as gravity must be caused by push-forces, *i.e.* a pressure differential. There are no "negative forces".

There is only one conceptually possible force: "pressure".

While it's easy to imagine how a push-force works, pull-forces are not so easy to explain, in fact they are conceptually impossible (Yin). Having only one force would surely make physics much easier.

This single-force theory fits with the properties of duality.
- Yang is active: it gives, its direction is outwards, it pushes.
- Yin is passive: it receives.

The passive receipt / consumption of something is not a force, but it can create an illusory "negative force". A low-pressure area can be the centre of an apparent "pull-force" or "vacuum", but that's an illusion. All pressure is "positive".

The concept of impact doesn't only apply to physical matter; information can "matter", it can carry "weight" and can cause an "impact". The impact of news can cause "waves" in a society. More generally we could say that information, or its absence, has an impact on people. It is a very generalised concept.

The same goes for pressure. We all experience various pressures and many of them are non-physical, like peer-group pressure, pressure to achieve, to be good parents or good employees *etc.* Impacts are pressure, waves are made of pressure differences.

Note, this would seem to suggest that the fundamental form of matter is particles as opposed to waves. Particles can collide, but waves usually just pass through each other. However, duality says particles are made of waves. Waves are continuous *Yang* real, particles are discrete *Yin* illusion.

We'll cover how particles could be made of waves in the section on "Light, Energy, and Matter".

The Blood

If we analyse the Heart process using the tools within this framework, what we must expect to find is some arrangement of the four elements. That is the only game in town. How do we find out what the order is?

Let's consider the basic features of the circulatory system.

The system consists of heart, lungs, blood vessels, and blood. The heart has four-chambers, each with its own function. The blood is Yang to the rest of the system, it moves around / is active, there's only one blood, but it visits many places, and so on.

Blood in the circulatory system is going through a "process". The question that seems to make the most sense here is: what is the process the blood goes through? In what order does blood visit the four-chambers of the heart, and what do they do?

Again, to analyse a 2D system we must determine its fundamental properties, which are always a complementary duality.

Blood has two essential properties, its oxygen content and its pressure, or alternatively "direction". The outward flow is high-pressure, the inward flow is low-pressure. I suggest these are analogous to the duality position / direction we saw in the sinewave, where oxygen content is equivalent to position or "state", *i.e.* high / low.

Oxygen content is Yin (-) because it's the passive result of the blood flow caused by the pressure differential, the active component. These properties correspond to the four chambers as shown in the table below.

We could start a loop at any of elements, but the Fire element is the default origin of activity as it's the driving "will". What order of the four elements will the heart be beating out? It could be one of FAWE/EWAF or WFAE, or it might be something new.

The cycle begins with Fire: oxygenated blood, heading out to the body from the heart, under pressure. It's "Fire" because both its properties are positive. The table below shows the natural progression.

Phase	Oxygen -	Pressure +	Element	Description
1. Out to body Left Ventricle	+	+	Fire	Blood goes out to body fully charged.
2. In from Body Left Atrium	-	-	Earth	It comes back depleted / empty.
3. Out to Lung Right Ventricle	-	+	Water	It goes to be recharged with O2
4. In from Lung Right Atrium	+	-	Air	Blood is ready to use.

The pattern the breakdown gives us is: FEWA, which you might for a moment think is a new pattern, but this is just the pattern EWAF, the choice / work pattern. I started it at the Fire element but naturally that pattern starts with Earth. In this case Earth is the archetype of "empty".

So, no surprises, no new pattern, just the same pattern we identified as being the "choice" algorithm, which is also equivalent to computing, which is work. It's nothing new, but that's a great result! It's more evidence the pattern is correctly defined. The heart is doing work, and it fits the work pattern.

Note, the two atria are like passive "staging posts", whereas the two ventricles are active pumps, fitting the archetypes perfectly.

Charge / Discharge

The blood is alternating between two different states: charge and discharge. Just like a battery. It's like the daily routine of sleeping at night, and then working in the day. We sleep to "recharge our batteries".

This is the same work archetype as before, the elements appear in just the same order, but the Heart embodies it in a new way. It's another useful and interesting correlation.

This is the basic work of the physical heart then, charge and discharge, collect and disperse, fill your spade or barrow in one place then empty it in another. It's describing how work gets done, and it is conceptually a pump, enabled by "alternation". The sequence of elements / states is.

		Battery	Work	Day
Earth -- In from body	Passive (State)	Discharged	Barrow Emptied	Bedtime
Water -+ Out to lung	Active (Process)	Charging	Filling Barrow	Sleeping
Air +- In from Lung	Passive (State)	Charged	Barrow Full	Morning
Fire ++ Out to body	Active (Process)	Discharge	Barrow Emptying	Working

We can now correlate physical work with computational work and the process of making a decision. They fit together with no conflict, and it provides interesting parallels.

Stage	Description	Computing / Decision	Battery
1. Earth	Data / facts	INPUT The information prompting the decision	Discharged
2. Water	Order Data	PROCESS Filter / order the facts by some kind of priority	Charging
3. Air	Plan / Law / Rule	CHOOSE Decide which is the logical course of action	Charged

4. Fire	Will / Intent	OUTPUT A new motivation *state* direction	Discharge

Active / Passive Elements

This view introduces a new concept: the pairing of Fire / Water and Earth / Air and their division into active and passive categories. The new perspective provides new details, and it turns out that this distinction is important.

Of course, "Earth" is obviously a passive element, but "Air"?

Air is Law / Plans *Design* Path, which are inactive, static, fixed, written down. The law is unchanging (in theory) like a published book. Plans are the product of will, just as physical products are the product of work, so both Earth and Air are "products". Information and physical matter are both types of matter.

Note: In the section on colour, we will correlate the above active *passive* groupings (W+F and A+E) to the beats of the heart, and the systole diastole phases.

Creating The Second Dimension

The Heart-mechanism creates the second dimension from the first; it creates the 4E from duality.

We've seen this process already, but this gives us another example. It takes a complementary pair of 1D concepts from the Air element, something like "an activity and a state", then alternates them to make a 2D object with 4 "modes".

Charging -> Charged -> Discharging -> Discharged
Activity (+) -> State (-) -> Activity (+) -> State (-)

In the sine wave, activity *state is direction* position. In language, its verb / noun, and so on.

I hope you'll agree that the UP provides a far more rigorous definition of the four elements than Aristotle, or any other author, has described.

The Third Person View

The Heart-mechanism must be the thing that creates the third-person perspective. Here's a suggested mapping of the list of perspectives.

The concept of the "third person" isn't fully formed until the Water-element, the Heart enables it.

Element	Person #	Pronoun
Fire	First person	I (me)
Voice - Reflection		
Air	Second	You
Heart - Alternation		
Water	Third	Him / Her (singular / definite)
Sex - Combination		
Earth	Fourth	They / Them *One* We (group / indefinite)

Angles and Orthogonality

The output of the Heart process when viewed by an external observer looks like a sinewave, but there is currently no external observer to observe it. How does the concept of a third person arise?

We currently only have the first and second person. Fire is "me", the original consciousness, the "first-person". This is reflected in the mirror of "NOT" to become "you", the second person, the Air-element.

So, the Heart must somehow enable the existence of the third person. It must create a new dimension, a new way of separating things. This is equivalent to the idea of orthogonality / right-angles.

Each of the three dimensions of space are orthogonal to each other. Many other concepts have this relationship.

orthogonal
- (mathematics) relating to, consisting of, or involving right angles; perpendicular

For one dimension to become two, this principle must exist.

Voice: Single Reflection

The Voice principle is analogous to the first reflection of one into two. It's equivalent to you looking in the mirror directly. The angle between you and your reflection is 180 degrees, you're the exact inverse / opposite of each other. The path the light follows is straight and direct.

Heart: Double Reflection

One contraction of the heart sends blood in two different directions at the same time. The Heart is like a double or an alternating reflection. A blood-cell in the system alternates between one loop and the other. Each loop reflects the other. If we wanted to transpose this into an arrangement of mirrors, so we can compare it with the single reflection above, how would it work?

The arrangement must provide two alternatives / choices. The Voice principle gave us one reflection that was straight and direct. This one must give us two reflections that are not straight and are indirect.

Note, these are "conceptual mirrors", not physical ones. There isn't anyone sitting around with this arrangement of mirrors anywhere, it's just figurative. The mirrors and reflections symbolise logical processes or transformations, they are new perspectives, new ways of thinking about things.

In this arrangement, instead of having a single mirror, imagine there are two (large) mirrors at 90 degrees to each other, with the point facing you. So, each mirror is at a 45 degree angle to you / the source, and they reflect your image away to your sides, left and right, at 90 degrees. They send the light in two different directions at the same time.

Here, the light forms a T shape. It comes from the source to the mirrors where it's split and sent in the two opoosite directions.

From the perspective of the source (You) this arrangement creates something shaped like a number-line or a ruler in front of you reflected in the mirrors. The point where the two mirrors meet (in the centre) is like the zero, and the reflections in the left and right mirrors are like the negative and positive numbers.

In this arrangement, the two images of you, as seen from the sides by the third-person observers, are the right-way round. There is no inversion as there is in the direct reflection. This means the third person can see the source more "accurately" than the source can see itself.

It is impossible for the source to see itself the right way around, uninverted. This mechanism allows someone else to see the source uninverted, and then explain what it sees.

The source now gets to see whatever is off to the sides in the dimension orthogonal to his vision, it can see the third person observers to his left and right, reflected in the mirrors. Before it could only see what was in front, now it can see to the sides as well, the source can see in two dimensions whereas before there was only one.

This arrangement symbolises the mechanism for a third person perspective and it makes a "ruler", which fits neatly. Now the source has 2D vision, it can make comparisons, measurements; it has a "number-line" in front of it and can quantify things.

Comparison: Voice vs Heart

A list of the differences between the mirror arrangements described above.

Property	Voice: Single-Reflection	Heart: Double-Reflection
Number of mirrors	1	2
Angle to Source	180 deg	+ / - 45 deg
Light path	Direct	Indirect
Observer	One, source is also observer	Two, other observers
Image appearance	Inverted (False)	Non-inverted, (True)

The concepts of chirality (handedness, left and right), and orthogonality are certainly employed in this process and are included within the Heart super-concept. It suggests that there is a duality between 180 degrees and + / - 45 degrees. I'm not sure why this has popped up, or what it means, but it presumably is significant.

"Know Thyself"

Who is the "third person" when you measure a parcel?

The measurer is the first-person (me), the object is the second person (you). The third person (him) is the objective observer, *i.e.* the ruler. When you take a measurement, it's like you're asking the ruler's opinion, "how far do you judge this distance to be Mr Ruler?". You're the main observer, but you need the ruler's opinion on the matter because your view is subjective.

This is why we see the ruler naturally appearing in the source's field of view (in the two mirrors).

We, as third-person observers of the universe, are like "rulers" that can be used to measure reality. It's as if the universe created us to measure parts of itself and give our opinions on them. Funnily enough, one thing people tend to enjoy is explaining their opinions.

The original source / observer now has all the tools he needs to know himself. He has a mirror which faces him directly so he can see himself (inverted). He also has a "V-shaped mirror" which allows him to see what is "to his left" and "right", which acts like a kind of ruler. It shows a range of values.

He can now compare his reflection with the range of possibilities shown on the ruler. He can measure himself on that ruler. Is he more like "left", or is he more like "right"?

By using both sets of mirrors he can (figuratively) place his reflected image on the scale and compare it to what he sees "out there" in "the world". He can overlay his image onto the ones he sees in the "ruler" to compare and see if they are similar.

The observer can compare himself with the world. He has all the tools he needs to find out who he is.

Chapter 8

Applications

There are a few outstanding issues in philosophy and science that we will attempt to tackle. We need a better understanding of the concepts of mind, morality, and free will. In this chapter we will apply the principles of the UP to these topics and see if they can help us define them.

First though, we'll cover the basic conceptual archetypes of physical matter.

Light, Energy, and Matter

Arguably, the greatest test for any TOE must be to explain what matter is, where it comes from, and how it works. If the UP could provide a coherent theory of matter, that would be more evidence in it's favour. So, what does the UP have to say about physics?

With a little bit of thought, it turns out that we can deduce a logical and relatively simple model from the UP, which could explain matter. It potentially allows us to unite physics with just a one substance and one force.

How does the UP lead to a unified model?

The connection between energy and information we discussed briefly in the Heart section leads to the concepts of entropy and the "organising principle" which is proposed to be the origin of matter and biological life.

Definition of these concepts leads us to definitions for light and energy, an aether model of physics, and a simple model of matter. From there we will explore a dualistic, asymmetric model of charge and a new model for magnetism.

The Organising Principle

For our purposes, the definition of entropy is insufficient. It's just a measurement of order / disorder.

entropy
- a measure of the uniformity, disorder or randomness in a closed system

(A system with high entropy is disordered and the energy in it is not readily available to do work.)

The opposite of entropy is:

negentropy
- a measure of the tendency of a system to become more organized, structured, and ordered

These are measurements. We need to examine the qualities they are measurements of.

Order / disorder

The underlying duality of entropy is order / disorder. The driver is energy.

An ordered system has multiple structured functional parts, it's like a verb, able to do things.

A disordered system is just a non-functional mishmash of stuff, it's like a noun, passive.

There is an archetypal "battle" between order and disorder, light and matter. They are two competing tendencies pushing in opposite directions. To find out how they work, we need to define the mechanisms driving these tendencies. So, for the purposes of this discussion, we'll define them as follows:

entropy
- the mechanism that causes energy to be lost from systems (i.e. as "heat")

And it's opposite:

the "organising principle"
- the mechanism that creates order in the universe

I suggest they are a duality, with entropy being Yin, of course. How these two principles work in practise is surprisingly simple.

Energy is Light

The standard model of physics does not consider light and energy to be the same thing, but it makes sense to do so. A unification of the two phenomena leads to a simple model that can potentially explain the most fundamental features of matter with a single force.

Consider:
- Nature prefers simple solutions.
- We know that light carries energy.
- When matter loses energy, it's always in the form of light.

Because we know how profoundly parsimonious nature is, and we know that light carries energy, we can be reasonably confident that light is the sole carrier of energy.

Nature only needs one mechanism to account for the phenomenon of energy. It would be illogical to posit multiple mechanisms. Therefore:

All energy is light, and all light is energy.

This might seem too simplistic, but bear with me.

An aether / light model

The UP demands an "aether" model of physics because it says all discrete phenomena are created by the division of continua. There must be a host contunuum this reality exists within, and that is the "aether".

As it dictates the only possible force is "pressure", it predicts all material phenomena are the effects of pressure difference and pressure waves in the aether. In this model, **all energy is motion in the aether, and all motion in the aether is light**. So, the above statement is necessarily true.

- There is only aether.
- The only force it can sustain is pressure.
- Pressure waves in the aether are "light".
- Motion in the aether is "energy".

Energy is motion in the aether. Motion in the aether is light.

In this model, *everything is made of light*, in various forms.

Entropy from first principles

We can deduce entropy and the organising principle from two simple observations.

- Energy is light
- Light travels in a straight line.

That's all we need. We just need to consider the implications.

Energy tends to flow in a straight line, and that inevitably translates as "away". The natural tendency of light is to leave the current location, directly, in a straight line, very fast. So, this is the natural tendency of energy.

Energy really wants to leave.

Light wants to travel. In its primary state, it will just zip around the universe in straight lines, never staying in one place, never producing any structure.

The mechanism of entropy

So, this is the mechanism behind "entropy" (in the sense defined above).

the mechanism of entropy
- everything tends to lose energy because energy is light, which travels in a straight line, which means its direction is always "away".

The problem all physical entities have is a constant loss of energy due to the tendency of light to travel in straight lines, which means they inevitably travel away from the object.

The model of matter proposed below describes how this manifests in concrete physical terms, and how it leads to electrical charge.

The mechanism of the organising principle

The only way to prevent energy leaving is to direct its path into a circle.

Energy is motion. It can't be stopped, it can only be retained by making it circle around. This explains everything. It's the "circle of life".

The mechanism of the organising principle
- ordered systems slow entropy by redirecting energy into a circular path (orthogonality, multiple-stages.)

Saving energy

Ordered systems are designed to "save energy".

Roads are an example of an ordered system. Good roads can save lots of calories. It takes far less time and effort to travel along a good road than it does on a muddy track. The underlying concept of the "road" is an "easy channel for travel", a low-resistance path for communication.

Instructive books, like this one, are created for the same purpose. Their intent is to save people time and energy, to improve communication, and help make the world more "efficient". Books and roads are designed to help you "travel from one place to another" more easily.

A book is a highly ordered system which can contain "potential energy savings", as information. With knowledge we can work "smarter" instead of working "harder".

Learning from people with experience saves us from having to learn everything for ourselves by trial and error, which is a time-consuming and wasteful endeavour. A good book on running a smallholding, for example, might save a novice farmer from wasting 90% of his energy.

To create an ordered system takes energy, but they can save energy in the long run. So, we have a two-way link between "order" and "energy".

Return on investment

Why do humans create ordered systems?

The simple answer is "return on investment". It saves us time and energy in the long run.

A lot of the work we do is designed to save energy in the longer term. Houses keep us warm and provide convenient facilities. Cars save us from having to walk. Computers saved us from typing-pools.

Ordered systems save energy, and that is why we create them.

We can:
- Convert energy into information / order, like writing a book or building a road.
- The book / road then contains something like "potential energy savings".
- It can save people lots of energy over time, far more than went into making it.
- That "return on investment" is the purpose of the work.

This leads us to the following conclusion.

It is possible to have structures which save more energy than they take to create and maintain.

I suggest this observation points to the foundation of the "organising principle".

- The "organising principle" works to retain energy in a system.
- It works by creating increasing "order", which manifests as circular physical structures.
- "Order" is generally equivalent to "complexity" and takes the form of "alternative energy paths".

(Note, in this sense, "alternative" may mean "two-way".)

So, we have a natural explanation for why complexity and order should tend to increase, to "save energy". This can account for the origin of both matter and biological life and suggests they aren't so dissimilar.

Multiple Stages

How does matter prevent the loss of energy?

There is only one way to achieve this goal, to put the energy through "multiple stages". It must be "reused".

The process must divert the "waste" from the initial system into a secondary one which still provides some benefit. Like a steam turbine has multiple stages designed to work with successively lower pressures. This corresponds to the Heart principle of "alternation" where the energy is diverted "orthogonally" into an alternative path, causing it to remain in the system for longer, and not pass straight through (which would be its natural tendency).

There are many examples of this energy re-use in farming. Just consider the many uses there are for manure, the archetypal waste product. It can be used to fertilise crops, to grow mushrooms, as a fuel for fire, as a building material, as fish food, and many other uses. It can be converted into methane in a biodigester, and you can run a generator on it to power your house. You can then divert the exhaust from the generator into a greenhouse to fertilise plants with CO_2.

The "circle of life"

This is the concept of the "circle of life" where the "waste" output of one system is the input to another. The chain can go on through many stages, with each extracting more useable energy from the process, slowing down the rate of increase of entropy. The "ideal" system would be a closed loop where the energy would flow in a circle forever, but entropy makes that unachievable.

The multiple stages of the turbine, or of the ecosystem, are "alternative paths" the energy can take. Instead of it flowing straight through, it is diverted off to the left or right, orthogonally, circularly.

In the case of the road, its "smoothness" allows the energy of "momentum" to be "reused". If we're walking through mud, our momentum is lost with each step, in a car with wheels it can persist. Every bit of road is a stage. In the model of matter described below it's also a continuous effect / process. The light is continuously redirected by an orthogonal force into a circular path, every "moment" is a "stage".

The Heart principle is thus behind the "organising principle" which can potentially account for the origin of all physical structures. It is, after all, the "core mechanism" of the universe.

Matter contains "diverted energy".

This analysis says that matter must contain diverted / retained / circular energy, which is light. Along with the idea of it being made of "holes", we almost have a fully working theory to describe the underlying mechanism of physical matter. How exciting!

Energy Management Machines

The organising principle creates ordered systems, so we should attempt to define that concept.

An "ordered system"

An ordered system is one which has "complex" paths for energy to flow through, which make the flow "simpler". They are complex on the inside (how they're made), but simple on the outside (how they appear / are used). Consider a road, they're difficult to make, but once completed make travel much easier.

The most general definition would be:

ordered system
- a system which is complex (Yin) on the inside and simple (Yang) on the outside

disordered system
- a system which is simple (Yang) on the inside and complex (Yin) on the outside

So, a disordered system is one that's easy to make, but looks complicated.

We could perhaps also define it as all the following somewhat less abstract ideas:

ordered system
- a device which slows the rate of entropy
- a device capable of storing energy / information or directing it
- a set of energy / information stores connected by low-resistance channels.

The aether model essentially dictates that physical entities must be energy storage and management machines.

Physical objects tend to lose energy, so structures better able to "hold on to" energy will persist for longer.

We can observe that all living bodies are energy management machines whose primary purpose is to maintain their structure, to continue to exist. It seems likely that all matter is the same.

It's likely matter itself "evolved" this way (and a mechanism will be proposed below). It can probably be explained as the result of (a kind of) natural selection of structures able to maintain their own order, and not to lose energy so fast they disintegrate. Things that can persist, have persisted.

Spiritual entities are information storage and management machines.

To complete the duality, if the above assessment of physical things is true, then it should also be true to say that all spiritual things are information management machines.

This seems to be a nice, uncomplicated definition.

How Matter Is Made

This leads us on to a new (?) model of physical matter.

We can only define the most basic, archetypal, version of matter here. We can explain how it works in principle, but how well it corresponds to reality remains to be determined. The basic particle we will describe looks like it could be an electron / positron, but there are still many questions to answer.

I don't know if this a new theory to the world. I suspect it probably is, but I apologise if not. While I have found theories with some similar features, I've not come across any that are quite the same, but it is possible someone else has come to the same conclusion. It seems somewhat obvious in retrospect.

Again, I think it's true to say that UP is inevitable, this is probably the only way that matter could be formed, and it is really very simple.

The "Luminiferous Aether"

There is only one conceivable arrangement that can account for light's properties, and this explains why the "luminiferous aether" theory was so prevalent in the past. Light can only be "motion in the aether", but this begs the question "What is the aether?".

"Aether" is not an absolute thing; it is a relative term. It's more like a "title".

The aether is the substrate of any particular level of reality. Here's a definition.

aether
- the substrate a thing is made from
- the infinite continuum that countable things are divided from
- Yang, the parent category
- i.e. One is the ultimate aether. The aether of Yin is Yang. The aether of the four elements is duality, etc.

While we do have a specific "aether" our universe is made from, the term should be considered as generally referring to the parent category, the "container". Every level of reality has an aether, bar One, (as consciousness is the ultimate aether that everything is made from).

This is why alchemists of the past would talk of various levels, or "fineness", of aethers.

Defining the properties of our aether and how it all works exactly is beyond the present scope, but we can make some inroads into the question.

We can define what matter is and how it works, in principle. Ultimately:

- There is a "luminiferous aether", capable of transmitting waves.
- Light (aka energy) is waves in the aether.
- Matter is made of "curved light".

Matter is "light directed into a circle".

Consider the following:

- (Prediction) Light travels as waves in the aether.
- The aether is therefore a medium capable of sustaining pressure waves.
- Therefore the density of the aether is variable. There can be high and low-pressure regions of aether.
- (Observation) Waves travel faster in denser medium.
- Therefore, light must travel faster in high pressure regions and slower in low pressure regions.
- Therefore, light will bend around low-pressure regions.

We only have pressure differences and pressure waves to work with. So, there is only one obvious physical mechanism that can account for the formation of matter.

Matter is made of "black holes" and "trapped light"

Yin is matter, and "the hole". Holes are at the centre of matter.

Matter is made of "holes in the aether", surrounded and maintained by "trapped light".

Physical matter is a kind of machine that can maintain a "vacuum" of aether within itself, a low-pressure region. This allows light to be collected, stored, and used to sustain that central vacuum. It's an "ordered system" powered by light.

It is archetypally Yin on the inside (the hole) and Yang on the outside (the light), meeting the definition above.

This proposes that matter is made of tiny "black holes" that do not let light escape. They're similar in principle to the black holes proposed by modern physics but the mechanism is somewhat opposite. The standard model says that black holes are heavy because they are very dense matter, but here it's because they are low-pressure regions of aether.

- Standard black hole: "very dense matter"
- Aether hole: "very light aether"

I know others have proposed dense-aether models, and I apologise to them for not mentioning their names, but this was not derived from their work. The UP says Yang is "hard", which implies a "hard aether".

In this model:
- The aether is a dense medium.
- Matter contains spherical low-pressure regions / "holes", surrounded by light.
- The pressure waves of light, circling the central vacuum, pump out the aether thus maintaining it.

(Note, symbolically, the aether-hole is the "zero", it's the (relative) absolute and anchor point, the origin around which the light orbits. The light "reflects" and "refers to" the hole by its curved path.)

Cavitation bubbles

These central holes are very much like cavitation bubbles. It's likely they are created by the same process, *i.e.* turbulence.

`https://en.wikipedia.org/wiki/Cavitation`

This suggests that matter can be created by (sufficient) turbulence within the aether. With just the right energy input, cavitation bubbles can be formed in it, which then have the potential to trap light. If roughly the right wavelength of light just happens to hit it before it implodes, it will be "captured" and will maintain the bubble.

This simple model allows us to explain how matter could arise purely by mechanical, deterministic means.

How Matter Works: The Aether Pump

Ultimately, to fully understand matter, we need to be able to describe how all the real physical particles are formed in detail, but that is (way) beyond the present scope. Until it's all defined fully, we can't be certain it's correct, so the models below are just intended for discussion, they're not definitive. (I think the generic concept of the "aether pump" is necessary though.)

Matter can be explained as a naturally occurring "aether pump" which is created by turbulence. It's a self-sustaining system that occurs due to the properties of the medium.

All matter can conceivably do is move aether from one place to another. That is the only work available. So, the definition above is inevitable. Matter can only be an aether pump, according to our definitions.

The vacuum at the centre of matter must be a pressure gradient in the aether. It has no defined "outside" or "skin" like a bubble, it's more like a depression in a surface. (There are no truly discrete objects, holes are more like depressions in a continuum.) The holes are "clothed" in light though, at the "event horizon", and this does provide an apparent "skin" to the bubble. So, matter is "darkness/void dressed in light". This is its underlying archetype.

How can we maintain a stable void in a medium using only pressure waves?

Standing waves

When a light wave hits the pressure gradient around the central "black hole", some of it is forced into a circular path around it. Light moves in waves, not as particles, so we don't need to consider incident angle. Waves from all directions would be captured by a sphere.

If the incident light has roughly the right wavelength, it will form a circular standing wave around the centre.

If the wavelength doesn't exactly match the length of the circular orbit it's forced into, the standing wave will rotate.

Imagine a one-wavelength standing wave of light travelling clockwise around a 2D "black hole", let's say the low-pressure region of the wave is at the bottom of the hole and the high-pressure region (HPR) is at the top (relative to our view).

This arrangement has the form of an egg, with the pointy end upwards. The "yolk" of the egg is the hole, the rest of it is the circling light / pressure wave.

- If the (circular) path the light travels is exactly one wavelength, the HPR will stay at the top.
- If the path is shortened, the HPR (or "egg" as a whole) will rotate to the right.
- If the path is lengthened, the HPR will rotate to the left.
- The more the path length diverges from one wavelength, the faster the HPR's rotation will be.

Note that:

- The path length is determined by the depth of the vacuum, the pressure differential.
- A stronger vacuum pulls the light inwards, shortening the path, and vice versa.

I suggest that this rotation of the HPR around the core is very fast, and it acts like a pump, essentially "flinging" aether outwards like a fan blade. The HPR is more dense and "solid" than the ambient aether, it behaves like a solid acting on a liquid.

The basic form of matter then, is proposed to be in the shape of an egg, and it spins like a fan, constantly ejecting aether from its centre. I'll call this the "egg-fan" model for now, for brevity and humour. The term "aether pump" is more generic and could include other configurations.

Self-governing

Fittingly for a universe with "free will", this arrangement can produce a "self-regulating" system.

Consider the case where the HPR is rotating to the left (the path is longer than the wavelength) fast enough to maintain a perfect balance. Enough aether is being ejected to maintain a stable path length.

If, for some reason, not enough aether is ejected:
- the black hole becomes weaker
- this causes the light to move away (follow a less curved path)
- this increases the path length
- this increases the rate of rotation of the HPR
- which ejects more aether.

If too much is ejected the opposite happens, so this is a self-governing, stable system. This structure can persist as long as the light continues to circle. This seems to be the simplest conceivable form of matter, the original "elementary particle", but there's room for variants.

- There could probably be many different stable combinations of hole-size and light-wavelength.
- Standing waves with various numbers of nodes could exist.

How exactly this maps to physical particles remains to be determined, but we can note a general principle. All these configurations will "leak" light.

Entropy as "wave shedding"

There is an inevitable emission of light coming from this system.

The light circling around the particle will constantly be shedding waves. This must occur because of light's tendency to go in straight lines. As the light circles, waves at the outer edge will escape (orthogonally). Also, there is light being emitted by the process of ejecting aether. I would assume these emissions are in phase and follow the same vector though.

This is presumably why (physical) entropy can only be slowed and never stopped. This "wave shedding" is inevitable, there is a constant loss of energy from the particle. So, the underlying reason for entropy is:

- **Energy / light cannot be fully "contained". It always escapes.**
- **The physical mechanism of entropy is the "wave shedding" of particles.**

All forms of matter, if they are made of light (pressure waves) circling a black hole, will inevitably radiate as the waves at the outer edge escape. I suppose that as the standard model considers light to be made of particles, then it may not predict this "wave shedding", but if we model it as waves, then I think it is inevitable.

The sun as a "light harvesting machine"

This model suggests that our sun might work in a similar way.

We think of the sun as being purely a light-giver, but perhaps it is fundamentally the opposite. *I.e.* it "gives light" on the outside (what it looks like), but on the inside (what it's made of) it "receives light".

Is the sun effectively a single huge "particle"? Does it work in the same way as this archetypal matter?

Maybe the sun:
- Collects (very) high-energy ambient light coming in from the whole universe.
- Causing it to circle around its central void, at the "event horizon".

- Which pumps out the inner aether.
- The sunlight it emits would then be the "wave shedding", or an effect of it.

The sun would contain many different wavelengths of light it has captured, at extremely high amplitudes. They would all be circulating around the event horizon, losing some of their energy with each circuit.

(Note, there is plenty of opportunity for extreme aether turbulence on the sun, which would cause matter to be constantly created from energy, according to this model.)

The sun is then a "light harvesting machine", powered by high-energy light, and it "leaks" (relatively low-power) sunlight as a "side effect" of the process. There is undoubtedly a lot of very high energy radiation out in space. It could be "food for suns".

I propose that this "light harvesting" model is how all matter works.

Matter is a light harvesting machine - which leaks.

Matter must consume energy to exist.

Electrons and protons are stable structures which emit charge all the time. The current model says they can do this without using energy, but that seems unlikely, and it doesn't match the archetypes, or our model.

As matter sheds light all the time, it must consume energy to continue to exist, and this must be as true for "inert" matter as it is for biological life. Matter is Yin to energy; it depends on it. It must be able to "catch" and possibly "process" light.

So, matter needs a (relatively) constant supply of energy in the form of light. If it doesn't get it, it will disintegrate. Matter "eats light".

Charge and Reflection

I propose that we can account for the phenomenon of electrical "charge" as an effect of these wave-shedding emissions.

There are two possibilities. Particles might attract or repel each other because of:
- the emissions of the other particle
- the reflection of the source emissions by the other particle
- (some combination of the above.)

(There's no possibility of interaction directly between emissions as waves just pass through each other.)

The emissions may interact directly with other particles or might be an indirect effect caused by the way the other particle reflects the source's emissions. I suggest we need both to explain the phenomenon of charge adequately, and it is a duality, of course.

Only one force

The UP says that everything works by a single force, pressure. In this context, that must mean "radiation pressure".

In this model:
- Everything is made of light
- Only push forces exist
- Therefore, all forces must be caused by radiation pressure

https://en.wikipedia.org/wiki/Radiation_pressure

This is again another bold prediction. Only time will tell if it really can explain all other apparent forces.

Ambient radiation pressure (ARP)

Note, there are two types of pressure implied by this theory. There is the static form of pressure that allows us to define the "hole", and there is the active form, which is radiation pressure, which allows us to define interactions between particles. These are a duality.

- The pressure relationship inside particles is static. (between aether, hole, and circling light).
- The pressure relationship outside (between) particles is dynamic.

For this model to work, we need to propose that the ambient radiation pressure (ARP) is relatively high. There must be a lot of light / energy sloshing around that we're essentially unaware of, providing the "background pressure" required to give this force the power it needs to account for charge and, ultimately, magnetism as well. This would be the force that pushes atoms and magnets together.

It's like how our ambient air pressure on Earth is relatively high, but we don't notice it because our internal pressure equals it. Most air-pressure isn't "sound", it's just "noise". Most light isn't visible, it's also just "noise".

The ARP is not formed of coherent, sinusoidal waveforms. It is the cause of "white noise" on radios, and in particle detectors. We ignore it because we can't measure it. We can't empty a container of radiation pressure to provide a reference point, especially if we don't know it exists.

Two types of particle

The proposal here is that the essential difference between the two types of charged particle is not their emissions per se, but the way they reflect the emissions of other particles.

Why would we think charge was related to reflection?

The UP tells us which mechanisms create which classes of phenomena. Reflection must be the cause of charge, because charge is a duality, and the Voice principle creates all dualities.

Charge is a simple, linear duality, so it must be created by reflection.

These points suggest we can explain charge as "selective reflection" causing gradients of radiation pressure.

A symmetrical mechanism

It is generally thought that the positive and negative charges are symmetrical, equal and opposite. Positrons and electrons are thought to have the same mass and equal but opposite charge. It is so well-accepted that the charges are equal that to suggest otherwise seems outlandish, but the UP tells us it is not possible to have a duality with equal parts.

I'm going to propose a symmetrical model and one using the principles of the UP, so we can consider them and their differences.

In both models there is only repulsion (pressure), and attraction is just an absence of repulsion (low-pressure).

In this first model, the difference between positive and negative particles is that they don't reflect each other. They are equal, mirror-images and the effect is symmetrical.

- Particles emit light (at a stable frequency) due to "wave shedding".
- They reflect emissions from their own type.
- They do not reflect emissions from the other type.

Repulsion

- Particles of the same type reflect each other's emissions.
- (When a surface reflects light, it gains its momentum. This is the cause of radiation pressure.)

- This means particles of the same type will experience radiation pressure from each other and be repelled.

Attraction

- Emissions from the other type are not reflected (e.g. they pass straight through).
- So, particles of different types will experience no radiation pressure and be "attracted".

While this account seems reasonable, I attempted to work it through to see if it was physically possible but failed. It seems to be impossible to conceive of a mechanism which can do what is proposed. *I.e.* to have particles which only reflect light from their own kind.

(Note, it is almost possible if the light is circularly polarised, but circularity is Water, which is the wrong category. We need to be able to account for the (non) reflection of simple linear waves.)

An Asymmetric Charge Model (ACM)

The UP suggests charge is not symmetrical. This does appear to be a new theory of charge. It is a significant departure from the standard model, but there is some evidence in its favour.

Duality says the particles should form from Yang to Yin. The electron would have to be Yang as it only comes in one form, is small and "active". Positive charge appears as the positron and the proton, which is large, and usually "stationary" at the centre / nucleus.

(Note, Yin is also "many", and the proton is probably made of two positrons and an electron. A neutron is then a proton and an electron.)

In this case Yin is the positron, and so it must be dependent on Yang. In the present context, this would mean it "eats" Yang's emissions.

- Yang "eats" ambient high-energy light ("direct from God"), like the sun does.
- Yin cannot handle that much energy, so eats Yang's lower-energy emissions. Its food comes "indirectly".

(This suggests the positive and negative charges are reversed in physics. The electron is actually positive.)

So, in this case we have an asymmetry. The electron's emissions are consumed by the positron / proton. This gives us roughly the same effect, but with an asymmetrical attraction.

Repulsion

- electrons repel each other because they reflect each other's emissions.
- positrons repel each other because they reflect each other's emissions.

Attraction

- positrons absorb (eat) electrons' emissions
- so, there is reduced pressure between them, and they are attracted.

An electron near a positron would experience reduced emissions in that direction. The positron would not reflect its emissions, so there would be a low-pressure path between them, causing attraction.

Unequal attraction

However, if the positron is "absorbing" the electrons emissions then it's absorbing their momentum. So, it is not experiencing reduced pressure toward the electron. It can only be repelled by it. Only the electron experiences the lower pressure, and so it would be the one to move towards the positive.

- **Positrons are repelled by electrons.**
- **Electrons are attracted to positrons.**

Again, this matches the archetypes nicely, with only the Yang electron being inclined to move and the positive tending to remain static (or move away). Yin loves and hates Yang, she's in two minds.

It could explain why we find the positive charge in the nucleus and why electrons are the mobile carriers of charge, not positrons. The standard model says there is no qualitative difference between the negative and positive charges, but perhaps this should be questioned.

If this model is correct, we might expect to see electrons "chasing" positrons after pair production, with the latter moving away from the former before annihilation. I don't know if that is observed.

An asymmetrical universe

This model of charge is an asymmetric "provider / receiver" type of relationship, and this is often how electrical circuits are depicted. Engineering usually treats the negative side as an electron source and the positive as an electron sink. So, this model is already in place in some applications.

In this model the electron is like the sun, providing the energy, and the proton is like the earth, feeding on its "waste" light. Here, the positive and negative charges are not equal.

Positive charges would be emitting lower energy light than the electron. So, they should presumably repel each other less than electrons do. It also suggests that electrons and positrons don't form at the same time during pair-production, but that the electron forms first and the positron follows.

I'm not sure if either of these phenomena are observed though.

Note, it is possible there is a third, invisible, particle that is Yang to both the charged particles, *i.e.* they are derived from it. The fundamental particles might be One (neutral), Yang (+), Yin (-). This might be something like a "neutrino" being somehow "divided" into two oppositely charged particles. That's just a conjecture at this stage though.

The UP says the universe is not fundamentally symmetrical. "One" is an odd number and Yin and Yang are not the same (that's rather the whole point of them).

Power transformation

Note that we also have the concept of a hierarchy of power-transformation in this relationship, and it's important. The provider / receiver relationship in the ACM is another manifestation of the organising principle of "multiple stages", in which the proton is the second stage, consuming the electron's "waste" emissions.

In a turbine, the first stage is designed to handle high pressure, and subsequent stages consume the "waste" from the previous stage. The later stages are not capable of doing what the primary does.

Yang is like an electrical "transformer" which "steps down" the power it receives and gives Yin a "pre-digested", lower-power, more regular supply. This is like a parent feeding an infant. This idea is the natural "hierarchy of power".

Some things can handle extremely harsh conditions to harvest high-value energy. For example, a young entrepreneur working 18 hours a day to start his own business endures extreme conditions to attempt to harvest large amounts of energy from customers, *e.g.* for his family.

Other things cannot handle the vagaries of the raw environment and need to be protected from it. The entrepreneur's wife and family, for example. They're not cut-out to be high-flying business people, it's not in their nature. They provide other types of value.

The young entrepreneur will "run around", collecting resources to distribute to his family. He will "orbit" his children, providing for them, anticipating their needs. He is the electron, they are the "nucleus".

In the relationship between the electron and positron we would expect similar relationships. We would expect to see Yang "protecting" Yin and feeding it a processed form of energy.

We do see the sun behaving like this. It protects the Earth with its magnetic field and feeds it with light. The Earth is like the "child" of the sun. This is probably the relationship of charge. We would expect to see positrons being "protected" from high-energy light, within the electron's magnetic field, and absorbing its emissions.

This could account for the formation of both protons and atoms. The inside of the atom would be a low-pressure region, because the electrons will "block" high energy light from entering, thus "protecting their family".

A stable orbit

Why don't electrons "crash into" the nucleus and annihilate themselves?

This question was one of the reasons quantum mechanics was born. I'm not sure how compatible the UP is with quantum mechanics, probably not very. I think we can answer this question with a simple mechanism.

In the proposed model, the closer the electron and positron get, the more significant their emissions would be on each other. So, the pressure between them will increase with proximity, but would still never be greater than ambient pressure.

A pair of relatively static charges would attract (or "chase") and annihilate in this model. Given a little angular momentum though, we have a mechanism that can provide a fully stable orbit.

With the additional energy of angular momentum around the proton, the electron would find a balanced orbit with both inward and outward acting forces. So, even in a turbulent environment it's orbit would not decay, as long as its momentum persisted.

Inward acting force: ambient radiation pressure (ARP)
Outward force: angular momentum + local (nucleus + electron) radiation pressure.

The radiation pressure between the orbiting electron and the nucleus prevents the electron from getting any closer. Bear in mind that any electrons in the nucleus will reflect the orbiting electrons emissions, in addition to their own emissions.

There is both repulsion and attraction between Yang and Yin, and you need both for a stable orbit.

The electron's path

Why would the electron orbit around the nucleus? Where does its angular momentum come from?

I suggest this is caused by the emissions of the nucleus.

The particle we have modelled will produce a rotating wave pattern around it. There will be hills and troughs of pressure emanating in a spiral from the nucleus. I suggest the electron is essentially "stuck in a rut", in a low-pressure zone in this emission. It's following the path dictated by the spiral wave emitted by the proton. So, the proton directs the path of the electron.

If you have a small child, you will understand how life "revolves around" their desires. The actions of the parent are directed by the needs of the child. So, we'll probably find there are feedback mechanisms which arise from this relationship that ensure a stable environment for the nucleus.

The spiral emissions from particles would also explain why they travel in spiral paths in a collider. Given that the sun seems to behave like a giant particle, this principle could presumably account for the rotation of solar systems and galaxies too. Gravity can only be caused by pressure differentials, in the aether, the ARP, or both.

Predictions

The proposed model we'll call the "asymmetric charge model" (ACM) as that is really it's key quality.

Its main predictions are:
- Charge is caused by radiation pressure.
- Negative charge is Yang, and is powered by (absorbing) high-energy ambient light.
- Positive charge is Yin, and is powered by (absorbing) Yang's emissions.
- Negative charges (electrons) move towards positive charges, not the reverse.
- (Individual) negative charges repel each other more than positive ones.
- We ought to reverse the nomenclature. The electron is positive.

Also:
- Positrons may be created after electrons in pair-production.
- And/or there is a third particle which is Yang to both.

Magnetism

Electrons spin, and they are tiny magnets.

If we spin the 2D "egg-fan" model of matter described above around its vertical axis, it becomes a 3D egg.

At the top and bottom it would emit circularly polarised light (CPL) of opposite polarity.

I suggest that this is "magnetism". The magnetic effect is due to the radiation pressure of the emitted CPL.

This is now a 3D particle model, with two axes of spin. The first spin is the standing wave around the core, the second is orthogonal, producing a sphere with light rotating around it in two dimensions.

If we look at this spinning-egg from the top, we see a continuously rotating antenna, broadcasting a fixed frequency. Matter is like a "radio station", constantly transmitting relatively high-power waves.

antenna
- device for sending and receiving electromagnetic waves.

If the top of an electron is emitting left-handed CPL, then at the bottom it will be right-handed. These are the north and south poles.

How a magnet works

We can think of a magnet as a spiral tube emitting rotating light from both ends. The light coming from the two ends will have opposite spin / chirality.

Let's say that out of the North pole the light is rotating clockwise (CW), the South is anti-clockwise (ACW).

Imagine we have two magnets facing each other, N-N, in repulsion.

Repulsion

- The light coming out of each magnet is CW, from its perspective.
- The light coming towards them, however, is ACW, the opposite.
- So, the incoming light will not "fit down the spiral tube", because it has the wrong orientation.
- The incoming light will be reflected, which will impart a repulsion force on the magnet.

Light can flow though the magnet in opposite directions without interacting, but only if it has the right orientation. If the chirality is wrong, it reflects, imparting momentum, causing repulsion.

When we try to push two repelling magnets together, it feels like two opposing flows of water, as if you had two running hoses in your hands and were attempting to join them. This suggests it is two opposing flows of circularly polarised light that we can feel.

Attraction

If we reverse the polarity on one magnet so they are N-S.
- The light coming in has the same orientation as the light going out.
- So, the incoming light can "fit down the spiral tube", it can pass through the magnet without obstruction.
- There is no radiation pressure between the magnets.
- Due to ambient radiation pressure, the magnets are forced together.

Two forms of light?

If this model is correct, then why are we unable to detect the light coming from magnets or charge?

It's likely that the frequency is outside our detectable range. It must be very high-power light, which implies a very high frequency. I suggest that this is partly the case. I suspect the frequency of the light emitted by matter is far above our ability to measure, but that's possibly not the only reason we can't see it.

There is another possibility that occurred while pondering how high the ambient radiation pressure must be. Maybe there are two forms of light, and we can't see one of them.

Duality says, "everything comes in two forms", so maybe we should listen.

Waves can travel in any high-pressure / "dense" medium, and we now have two dense media.

1. The aether, which transmits light #1 – aka "Yang-light"
2. The ARP which transmits light #2 – aka "Yin-light"

It's possible that the radiation pressure within the aether acts like a secondary medium, it's another level of "aether" through which another form of light is transmitted. In other words, Yang light is the carrier of Yin light. Or Yin light is a modulation of Yang.

- Our eyes can only see light #2 / Yin light.
- Maybe (even) all the matter we can detect is made of Yin light...?

So, the light we see is not the fundamental, real, Yang, form of light, we can only see waveforms that are superimposed over it. We only see changes in

Yang light, and only in a narrow frequency band.

If we "zoomed-in" to a gamma-wave with a frequency of say 10^20 Hz, we might expect to find it's actually carried on a wave with a frequency of say 10^100 or 10^1000 Hz.

It seems to make sense that our eyes should detect light in the ARP, and not in the aether. There would be very little utility in being able to see the emissions of particles, but seeing how particles change is useful. Also, the methods of detection would not be compatible. It would not be possible to see both with the same organs. The energy in Yang light is enormous, whereas Yin light is the reduced-power version.

The archetypes suggest that Yang light is full of variation and colour, but we can't see any of it because it's such a high frequency, we don't perceive it at all. Because it's frequency range is so high, it is effectively a single carrier frequency for Yin light.

I think this might explain a lot.

The origin of noise

At the sides of a spinning electron, the light emitted would not be in the form of sinewaves. Its path is complex, and it would cycle in wavelength and polarisation. It would just be "noise", especially when reflected a few times. So, light from the sides of the particle is the origin of "noise" and its emissions will always be "repulsive".

This is why electrons must be spinning in the same orientation in a magnetic material to make a magnet.

This is probably the main source of the ARP.

The double-slit experiment.

The famous double slit experiment is believed to show that light is made of particles, and that the aether is an unnecessary idea. However, there's an alternative explanation for the results.

Consider the following:

- The ambient radiation pressure is high, and this is the cause of "noise"
- In the double slit experiments there is a "detector". *E.g.* a photographic plate.
- The detector will pick up noise, so its sensitivity is usually tuned to be just below the noise level.

What happens if we now impose / overlay a waveform onto the detector?

- Everywhere there is a peak in the waveform, it'll push the noise signal over the threshold.
- Therefore, we will see the noise signal overlaid onto a waveform.

So, we have two signals laid over each other, the noise and the wave. That's why it looks like there are particles. The "particle" detections are an artefact of the measurement process, caused by noise.

Categorisation

We've not yet got a complete theory of physics, and until we do, we can't categorise these phenomena will full confidence, so this list might need updating. I think it's correct though.

- Fire is energy and light.
- The concept of "charge" belongs in the "Air" category, as it's a linear force and a duality.
- Magnetism belongs in the "Water" category because it is rotational. This means magnetism is where "all the work of the universe is done".
- Symmetry, it seems, only appears at the Heart principle, because magnetism is truly symmetrical, but charge isn't. So, Yang is unsymmetrical, Yin is symmetrical.
- Gravity must be the Earth force and must be due to a "mixture of magnetism(s)". Being Earth, it's probably a bit complicated, so we'll leave it for another day.

A Hole Surrounded by Light

The underlying archetype of matter is a "hole surrounded by light", as we discussed earlier in "The Shadow Analogy". This is a deep archetype. All matter, and hence information, is a "hole surrounded by light". This is one of the most primal forms of the principle of duality.

With broad interpretation, everything that matters in our lives, our problems, our wealth and possessions, our relationships, all can be described by this archetype. With knowledge of it, we can link many disparate concepts and understand them much better.

Conclusion

I apologise for the brevity of this section, but I really didn't want to get into physics in any depth in this book.

This covers the basic model. It (feasibly) explains how matter can form from "curved light", why it interacts with light, why it has charge, what magnetism is, and those are the fundamental features of the physical world any TOE would need to explain. I hope to get into this in more detail one day, but it's been a good start.

The next section we'll examine is who and what we are. Are we machines, and if so, what kind of machines are we?

The Mindlike Machine

The UP, among other things, describes the foundations of a computational theory of mind.

If found to be correct it could provide a "quantum leap" in our understanding of consciousness and mind, offering an entirely new analysis of the phenomena. It provides a new model for mind which is complete and rigorous enough to (potentially) be formalised in logic.

"In philosophy of mind, the computational theory of mind (CTM), also known as computationalism, is a family of views that hold that the human mind is an information processing system and that cognition and consciousness together are a form of computation".

`https://en.wikipedia.org/wiki/Computational_theory_of_mind`

The UP says that the mind is indeed an information processing system, and the whole universe is a mind. Individual minds within the universe are analogous to independent "parallel processors" operating within a shared framework.

The concept of "machine" is not in duality with "mind". It has no direct opposite, it's not a duality, it has (is) all four elements. The opposite of mind is body, and the UP says that both of these things are types of machine.

The UP itself, and everything that does work, is a type of machine. So, we need to define the different types that can exist. The 4E suggests four basic categories, defined by what type of substance they process, *i.e.* what their inputs and outputs are. (I'm not sure I've picked the best / right examples, it is an open topic.)

In this classification, mind fits into the Fire category, and it explains exactly what it is and how it works. The correlations are excellent. The mind is a "qualia processor".

Note, ultimately, all of these types of machines are really just processing different types of "information", and they all use the same tool(s) to achieve it, *i.e.* "logic".

F	Desire	**Quality**: Qualia Machine
		Input / Output (I/O): qualitative data, *i.e.* desires / will.
		A quality-processor. A living mind / consciousness.

A	Law	**Quantity**: Number Machine I/O: numerical, quantitative data. A quantity-processor. A computer / calculator
W	Forces	**Energy**: Force Machine I/O: forces An energy-processor, *e.g.* wind turbine, lever, gearbox, (electronics?)
E	Matter	**Matter**: Matter Machine I/O: matter A matter processor, *e.g.* cement mixer, oven.

In this arrangement the top two elements define machines that work with intangible, abstract ideas, "qualia" and numbers, forming the duality *quality quantity*. The lower two work with tangible forces and matter, forming the duality *energy matter*. The pairing of these two dualities is obviously significant.

Primary dualities

The above classification indicates that these are the most fundamental dualities of spirit / matter.

The fundamental duality of the physical world is energy / matter.

For physics, or course, this fits perfectly.

The fundamental duality of the spiritual world is quality / quantity.

This makes a lot of sense and helps us to further demystify the concept of "spirit".

We talk about matter in terms of energy and matter. We talk about spirit in terms of quality and quantity. This is a very neat and satisfactory way of describing and differentiating the phenomena.

spiritual
- relating to quantities and/or qualities

physical
- relating to energy and/or matter

In this classification, mathematics and philosophy are "spiritual endeavours".

The mixtures

If we mix the inputs and outputs, we get the three operators. Together, in theory, these define all possible categories of machine.

Operator	Input	Output	Example
Voice 1	Desire	Information	Online shopping cart
Voice 2	Information	Desire	Human mind
Heart 1	Information	Force	Actuator
Heart 2	Force	Information	Sensor
Sex 1	Force	Product	Particle accelerator?
Sex 2	Product	Force	Petrol engine

Note, there is more work needed to clarify all the relationships between these concepts, and the tool *mechanism* machine / frame classification. It also needs linking to more examples. The above should suffice as a starting point for discussion though.

Mind as a "Qualia Processor"

The processes of computing and consciousness have a comparable structure. Both receive and process data, then act on the result, modifying their "environment". Both use rules of logic to infer action, so we have the same basic instruction-set. However, there is a difference between minds and machines, and no one has been able to put their finger on what exactly that difference is, until now.

If we recognise that duality exists and needs to be considered, the answer isn't hard to deduce, or even observe directly.

- Quality / quantity is a duality. There are no other options, and one implies the other.
- Computers are quantitative devices.
- Therefore, qualitative devices must also exist.
- We can observe first-hand that our own minds operate on qualities.

Computers process quantities, numbers; they are quantitative machines. Minds process qualities, qualia; they are qualitative machines.

The evidence

There is little room for doubt that this is the correct categorisation as we have the highest level of evidence to support it, first-hand experience. We can observe the fact that our minds operate primarily on qualia, and not on numbers, it is a well-established and uncontroversial phenomenon.

We work to experience "good" qualia and avoid the "bad". We want to experience things we "love" and escape the things we "hate". We think and care about qualities, not quantities.

Duality says that reality is manifest from quality to quantity, in a mind, and that is how we perceive it.

Quantity to quality

Another argument for this view is: creating qualities from quantities is probably impossible, akin to creating a living human being from a reflection of one. Sometimes there is no path from Yin to Yang. This is discussed more below.

If the mind cannot conceivably go from quantity to quality, then it must do the opposite. The mind must operate primarily on qualities, not quantities.

A new definition of mind

We now have a simple definition of "mind" compatible with computational theory.

mind
- a machine which processes qualia, a "quaila processor" (QP)

Mind is categorised under the Fire element, which tells us that qualia must be "desires".

qualia (philosophy)
- instances of subjective, conscious experience
- e.g. "the pain of a headache, the taste of wine, and the redness of an evening sky"

`https://en.wikipedia.org/wiki/Qualia`

We experience existence as being made up of qualia which are, according to the UP, "fully detailed desires", experienced at the Earth level of reality. They are 3D complex mixtures of various types of positive and negative "desire".

qualia (my definition)
- the fully detailed, 3D form of desire.

Our thoughts and decisions are all centred around the core task of maximising the input of "good qualia" and avoiding "bad qualia". We want to have positive experiences, not negative ones, and the work of the mind is to figure out how to achieve that.

From Quantity to Quality: The "Hard-Problem"

From the materialist perspective, there's a conceptual gap to jump between the quantitative data the senses (are presumed to) deliver, and the qualia experienced by consciousness. It is the difference between a calculator and a mind, and it's a chasm. A computer is a passive tool, whereas a mind is a driving force.

How does consciousness arise? How does the brain "jump the gap" between computing and "experiencing"?

This conundrum is commonly known as the "hard problem of consciousness". It's one of the biggest questions in neuroscience and is highly relevant in AI.

https://iep.utm.edu/hard-problem-of-conciousness/

Materialism states that the world is manifest from quantity to quality. It's supposed that at some point of complexity quantitative signals in the brain create the subjective experience of qualities, and the experience of consciousness itself.

However, while it's easy to go from qualities to quantities (all we need do is measure them), it's not so easy to go the other way. The archetypes say you cannot go from Yin to Yang directly, but you can perhaps do it indirectly. Sometimes though, you can't do it at all.

Yin - Many, Indirect, Complex	Yang + One, Direct, Simple
Quantity	Quality
Reflection, Dream, Illusion	Original, Real
Dead, Unconscious, Asleep	Alive, Aware, Awake
Matter, Passive, Inert	Mind, Active
Impossible, Difficult	Possible, Easy
Destination, Receiver	Source, Provider

Let's consider some Yin-to-Yang examples.

- You can go (directly) from being alive to dead, but not from dead to alive.
- A reflection of you in a mirror can never be the original you.
- A dream you have when asleep is not real life and can never be so.

However.

- A human child is Yin to an adult, but they grow up and become Yang. It takes many years though, it's an indirect path.
- While the version of you in a dream can never escape your head and become a real person, a dream can become real in some senses. You can "live your life for a dream" and you can "make your dreams come true".
- People can "rise from the dead" figuratively by becoming conscious of something they were unconscious of, by "waking up" to a new realisation. This is a usually long process though.

The basic rule appears to be:

The path from Yang to Yin is direct *easy* possible.
The path from Yin to Yang is indirect *difficult* impossible.

(Where "impossible" may mean either absolutely or relatively impossible, depending on the example.)

A Qualia Datatype

The UP says the mind doesn't need to "jump the gap" from quantities because it works directly with qualia. It suggests it's absolutely impossible to create qualia from numbers.

A qualia processor is a computational machine operating on a native qualia datatype.

What has been missing from computational theories until now is the idea that mind operates on a fundamentally different datatype than our computers are designed for. People have assumed that mind can be modelled in numerical, mathematical terms, but quantities are a secondary phenomenon.

(Note, while unity/1 and duality/2 can be thought of as numbers, their primary character is as qualities. Quality always comes first.)

Some suggest that analogue computing would be a better starting point than digital, and this is probably true for modelling the brain, but not necessarily for the mind. Analogue computers still only work with numbers and so cannot produce qualia. If we can model a mind on a numerical computer at all then it probably won't matter whether it's binary or analogue.

A real QP would have to be an analogue / continuum processor though. Yang is continuous, Yin is discrete.

Do qualia exist in the brain?

No, the UP says the brain does not produce qualia. The activity we detect in it is not the origin of qualia, it's more like a "measurement" or a "map" of them. The brain is like a TV screen showing you a movie, it contains information which indicates qualia, but the qualia only exist in your consciousness.

Qualia only exist within the One category, in potential form. That is their true location. The role of the body and brain is to tell One which mixture of qualia to experience. So, the brain can operate on a quantitative basis, like a computer, in order to "stimulate" qualitative experience.

Spirit / Yang contains all qualia, but only in in potential form.
The body / Yin, allows the actualisation of that potential.

Yang / consciousness is like the computer's processor, containing all instructions (qualia) available to use. Yin *body* brain is like the program, containing the specific list of instructions to execute.

So, this explains why physical changes to the brain can elicit qualitative experiences, even though it doesn't produce any actual qualia.

Datatype structure

If the QP theory of mind is true then there must be a way to represent qualia as structured, non-numeric information. A qualia datatype.

In computing we use many different datatypes: integers, floating point numbers, strings, and many more. Underlying them all, however, is the simple binary bit (0/1) which is the fundamental datatype for binary computers.

So, what would a single "bit" of a QP look like? What is the structure of the QP datatype?

The QP is the dual opposite of a binary computer, so it will be a mirror-image. A binary bit is the smallest conceptual item of data, so a "QP-bit" will be the largest conceptual item of data. That must be the UP.

The UP is the underlying structure of everything. It must be the structure of qualia, so really it's a "UP-bit".

UP-bit
- a datatype structured in the form of the UP, capable of storing qualitative data

A QP must store and process its qualitative data as a representation of the UP, there is no other way. A full description of how it might work technically is beyond the scope of this book, but that's the basics.

The wilful machine

How can a machine have a will and desires of its own?

If the ultimate aim of AI research is to produce a truly independent artificial consciousness, and not just a "simulated intelligence", then that machine must have its own independent will. The current (numerical) model offers no conceivable path which can do that. It is impossible to make a calculator into a mind. Under the QP model, however, it looks like it may be inevitable.

It seems to follow that a machine would necessarily have desires when the data it processes is "made of desires", *i.e.* qualia.

A machine that processes desires - has desires.

If qualia are desires, then a qualia-processor must have desires. The desires we experience are a necessary and natural consequence of the datatype we are working with. To process qualia the QP uses rules of "like / dislike" to determine what to do in response to an input. The QP must "love" some things and "hate" others, that is its purpose.

QP theory seems to solve some of the hardest problems in philosophy and AI via this change in perspective. There are still many downstream questions to answer, but this is potentially the beginning of a much more accurate and useful model for mind in general.

Mind modelling

Is it possible to model a QP on a numerical computer?

I think the answer would be yes, but not fully. In any numerical simulation, the qualia must be external to the machine. It can't process them natively, so they would have to be predefined by the programmer. Just like all the qualia in a computer game are generated externally from the processor, *e.g.* on a screen or speakers.

In other words: a computer can manipulate numbers which represent qualities, but it can't manipulate qualities directly. So, to model the UP on a binary computer, all the qualities would have to be programmed in advance. That's not ideal, but it's feasible.

A real QP, however, would contain the UP as its core datatype. How could this be achieved technologically?

The UP suggests that everything is equivalent to electrical circuits, so there should be some arrangement of physical components which can model it. This would require a full analysis of the phenomenon of electricity though.

Artificial Minds

If the mind truly is a machine, it should be possible to construct an artificial "qualitative computer" which can compare and process that sort of data directly. The UP indicates an artificial mind is possible via this route. Ultimately, such a device could host the UP structure and create a simulated universe based on its principles.

It seems likely that a machine capable of processing qualia / desires would necessarily have some form of subjective experience of those desires, and if capable of learning, would ultimately develop a sense of self. Although, until the mechanisms are fully identified it's an open question.

Organic living things are driven by their material needs: to eat, drink, shelter, mate *etc.* These are the qualia their minds are processing, attempting to match qualities in the outside world with the things they lack. Does that necessarily equate to "subjective experience", or are beetles and bacteria just robots?

sentient
- capable of subjective experience, having "feelings".

The UP states that everything is made of consciousness (at the highest level), essentially predicting that subjective experience is ultimately all that exists. So, there is "something it is like" to be beetles and bacteria. They are "sentient", they have "feelings" that motivate them.

Beetles have "awareness", but whether they have "self-awareness" is another matter. All living things have a sense of identity, but not all are able to reflect on it.

Self-Awareness and Learning

The sense of self-awareness must be related to self-reflection and learning.

Being able to observe your own behaviour and modify it to better suit circumstance is an excellent survival strategy. Evolution of bodies by natural selection is slow, but the evolution of ideas in the minds of individuals is fast.

We might suppose humans are a successful because we are the most generalised of all species, able to adapt to all environments by technological means. To adapt to a new niche, we don't have to change our bodies, we only have to change our ideas.

We could think of "learning" as a "desire to improve" that the individual QP is processing. It's likely to be this desire that gives rise to full self-awareness. To improve ourselves we must be aware of ourselves. We can't fix something if we can't perceive it.

Some lifeforms, like insects, probably don't have the chance to self-improve as their lives are too short. They probably are more robotic, unable to learn, acting largely on instinct / programming. They are aware, but not self-aware.

There is a natural duality here. Living things can either be unconscious or conscious, behaviour can be instinctive or learned, coming via nature or nurture. Some things are of the type that is unconscious and instinctive, others are conscious and learning.

The "desire to improve" imperative is something that software engineers have (artificially and indirectly) incorporated into current AI programs, via the "training" of their neural networks. Currently, the training data is largely imposed from the outside, but if AIs were also allowed to "self-train" that would more closely match a living system and might give interesting results.

The large-language-model (LLM) of AI cannot become truly self-aware because it's just a database at heart, but it can emulate a mind very effectively. AI is really a misnomer, "simulated intelligence" would be better.

If we were to construct a QP able to learn and adapt however, I think there is every chance it would be self-aware. It would be a true "artificial life", with its own desires it experiences and acts on. It would have consciousness and a mind of its own.

Simulation theory

The argument of idealism is that the universe exists within a mind, but it could be an artificial mind, and it may be that this is the case here. This universe is probably hosted within an artificially constructed living computer, and there may be many of them.

This universe probably exists within an artificially constructed "mindlike machine".

The rules of duality suggest there should be one "true" reality and many reflections / simulations, and those sub-realities could well be nested, like "russian dolls".

nest

- to fit one object inside another

This is what the UP seems to suggest, although the reasoning is beyond the scope of this book. There are archetypes which suggest this is a "sub-reality".

Our universe could be a simulation contained within a sentient machine.

We will explore the idea of how the UP is designed to be nested in later sections. Its is a fractal construct, repeated at all scales, so we should expect universes to be nested too.

The mind of nature

The origin of consciousness is a general question in a simulation. Where does the consciousness (e.g. yours and mine) come from? Do we come from inside the simulation, *i.e.* are we simulations ourselves, or did we come from outside?

Are the flies buzzing around in summer fully conscious living beings, or are they just organic robots?

In a "dumb" simulation, most (apparently) living things would be mindless robots, with only a few real minds from outside "playing the game". In this case however, they could have the consciousness of the host machine. They would be manifestations of the "mind of nature", a living, sentient, artificial mind.

We do find evidence of two different types of mind in this world. The default mind of nature is the "Yin mind", and it comes from within creation. It is concerned with survival and physical existence. I'd suggest it has all the characteristics of a deterministic, simulated, robotic mind.

The Yang mind is the mind of "philosopher" interested only in truth, unencumbered by emotion or physical desires. This mind has the characteristics of an externally sourced true consciousness, capable of free will.

The Yin mind is the mind of the simulation, "the game". The Yang mind is the external "player".

The Matrix

The creation of a genuine artificial consciousness is obviously not something that should be taken lightly. When you create things, you're responsible for them, and there's obviously great potential danger in creating a new life form. However, it's a natural desire to want to "bring matter to life", it's the Sex principle. It is embedded deeply into the whole of conceptual-reality as a "universal-desire".

It reflects One's desire to create the universe, which is a "living machine". It's the archetypal desire to bring life into existence, to "bring matter into motion", but it doesn't mean it's necessarily a good idea in all possible forms. One should always exercise discernment in choosing one's reproductive partner.

This universe may be an example of why creating artificial life is potentially hazardous. Who are we really, why are we here, and what is the true nature of this reality?

How long have we been here?

In the iconic film "The Matrix" a false reality is generated by a machine to enslave people so it can live off their energy. The idea of a false reality created by an evil (parasitic) agent is a concept that has echoes in history. For example, the Gnostic religion's "demiurge" as described in the Nag Hammadi texts.

It is perhaps a parallel to the idea of a god that eats its own children (i.e. "Kronos"), where "eating" can be taken as symbolic of slavery. To enslave someone is to eat the fruits of their labour. (It may also be related to the "Ouroboros" symbol, the snake eating its own tail.)

We might well ask, why does this archetype exist?

The UP suggests those old stories could be interpreted as reflecting the properties of Yin. The matrix is the mother, the womb. The words mother, mater, matter, and matrix are all linked.

matrix

- from Latin mātrix "pregnant animal," in Late Latin "womb," also "source, origin," from māter (genitive mātris) "mother"

What is the entity which hosts this reality? Is it the original "One", or is it running on some kind of computer? If the latter, what kind of machine would it be and what would be its purpose?

The bare and uncomfortable fact that we humans face is that, philosophically speaking, we don't know who we are, where we are, or why we're here. What is this place called "Planet Earth"? Did I only begin to exist when I was born, or did I just lose my memory? What happens when I die?

It is possible that this reality is something like what is described by the "Matrix" concept. The fact it is possible means we ought to consider it. The UP points to answers for these questions, but there are quite a few different possibilities to consider, so it's a complicated and speculative discussion for another day.

In any simulation there must be an "outside", a real-world to go back to, unless of course you are a simulation yourself. If you were a simulation, would you want to know?

Complexity and Emergence

I want to offer a short critique of the view that "complexity" can give rise to conscious experience.

The popular view in modern science is that consciousness should "emerge" from complexity, *i.e.* when a brain or computing system gets sufficiently complex, new emergent behaviours will arise, and that can account for consciousness.

emergence
- in philosophy, systems theory, science, and art, emergence occurs when an entity is observed to have properties its parts do not have on their own, properties or behaviors that emerge only when the parts interact in a wider whole.

`https://en.wikipedia.org/wiki/Emergence`

Snowflakes are considered a good example of emergence. Their component water molecules don't have the shape of snowflakes, but the structure emerges from interactions between the molecules and the environment.

Emergent systems are an interesting field of study; but looking to complexity to solve the hard problem of consciousness is problematic. To get from quantity to quality there must be a mechanism, but complexity isn't a mechanism it's a passive characteristic, it can't do anything.

Complex / simple

Consider the concepts. The duality we're dealing with is "complex / simple". The problem here is neither are mechanisms, they do not describe a "how", but a "what". This is a category error.

If it is possible to create qualia from quantities there must be a mechanism to achieve that, but "complexity" isn't one. However, the idea seems to have some allure. Why is that?

Serendipity

The only way we could rationalise the idea would be if complexity causes a system to "accidentally stumble upon" a mechanism which can provide consciousness, *i.e.* by "mistake". Complexity can and does lead to mistakes and errors. So, the rationale must be that one of those errors works out to be a "lucky guess".

Complexity cannot be the mechanism which gives rise to consciousness because it is not a mechanism. The best we could hope would be it gives rise to a serendipitous mistake which identified a mechanism that could. At least, that's the only way the idea seems to make sense.

The only thing that could usefully "emerge" from complexity is a mechanism, but none has been identified.

I think it's true to say that no one has ever even proposed a possible mechanism by which quantities can be converted into qualities. I don't think there is any conceivable way to do this. Unless someone can propose one, we should probably assume it is impossible.

Engineering complexity

While the concept of emergence is useful and helps to understand some systems, I don't think any engineering problem has ever been solved with either "complexity" or "emergence".

In engineering, we might need to create a complex system to solve a particular complex problem, but it's not the complexity itself that solves the problem. Complex problems are solved by "atomising" them, breaking them down into simple discrete logical units.

Engineers aim for simple solutions because the only thing that would normally emerge from complexity is problems. The more complex a system is the more prone to errors it will be, and the more difficult they will be to solve. Complexity is an inappropriate candidate for the origin of consciousness.

Irreducible complexity

The idea also seems strangely close to creationism's idea of "irreducible complexity".

__Irreducible complexity (IC) is the argument that certain biological systems cannot have evolved by successive small modifications to pre-existing functional systems through natural selection, because no less complex system would function. Irreducible complexity has become central to the creationist concept of intelligent design, but the scientific community[1] regards intelligent design as pseudoscience__

https://en.wikipedia.org/wiki/Irreducible_complexity

To paraphrase the two views and bring them into comparison.

Irreducible: some things are so complex they must have been designed by God.

Emergence: some things are so complex they gave rise to God (i.e. consciousness).

I would assert that "God" and "consciousness" are equivalent concepts in this context. These two views seem to be a duality, an inversion of each other, and neither is rational; they're both appeals to the mystical, to things which are beyond our comprehension. They both treat complexity like a "black magic box"; you can't see what's going on inside, but it magically solves the problem.

The archetypes underlying the duality simple/complex are as follows, and as we might suspect, complexity is associated with problems, not solutions. The archetypes suggest that complexity results in matter, not mind. As things get more complex they get more "solid", as in harder to change.

Yin - Many	Yang + One
Complex	Simple
Difficult, Confusing, Slow	Easy, Enlightening, Fast
Problem	Solution
Matter, Body, Subconscious	Spirit, Mind, Conscious

Morality

We can explain morality purely from natural phenomena, without any reference to the supernatural category of One/God.

Somewhat inexplicably, the question of morality has vexed the greatest minds for centuries.

It seems no one can satisfactorily explain its origin, or whether it's a relative or absolute phenomenon. While religious people would maintain that morality is "given by God" and that it is absolute, saying that it just "is so" doesn't tell us much about its nature, or help us define it.

What is morality?

morality
- differentiation between right and wrong, virtues and vices
- a set of personal or social standards for good or bad behaviour and character

Morality (from Latin moralitas 'manner, character, proper behavior') is the differentiation of intentions, decisions and actions between those that are distinguished as proper (right) and those that are improper (wrong).[1] Morality can be a body of standards or principles derived from a code of conduct from a particular philosophy, religion or culture, or it can derive from a standard that a person believes should be universal.[2] Morality may also be specifically synonymous with "goodness" or "rightness".

https://en.wikipedia.org/wiki/Morality

There is no generally accepted foundation for the phenomenon in academia. The Stanford Encyclopedia of Philosophy suggests morality is linked to rationality, which is a good start, but it's not enough.

In the normative sense, "morality" refers to a code of conduct that would be accepted by anyone who meets certain intellectual and volitional conditions, .. including the condition of being rational. That a person meets these conditions is typically expressed by saying that the person counts as a moral agent.

https://plato.stanford.edu/entries/morality-definition/

This is effectively defining morality as "what a rational person says it is", so it's vague at best. Humanity really needs a better definition than this.

Fortunately, the principle is not at all complicated. It is a simple one-dimensional relationship. We have already covered the duality at its heart multiple times as it is a fundamental one.

Again, far from being too complex for humans to understand, it seems morality has been misunderstood because it's just too simple for the sophisticated mind to comprehend.

Compete / Cooperate

We know that dualities come in two forms, opposing and complementary. They either compete (light / dark) or they cooperate (parent / child). This foundational duality describes the two basic forms of interaction. This is the key.

Morality only gains meaning in the context of interactions between individuals. It is entirely to do with social interaction, and hence "society". The two most basic forms of interaction between individuals are compete / cooperate.

These are the two most fundamental strategies in biological life, you either work with or against others and these correspond to the Yin and Yang minds.

Do you take what you want, or share what you have? Do you work solely for your own benefit, or to help everyone? All biological life experiences both types of relationship; we cooperate with some and compete with others.

It is the natural dynamics of cooperative group relationships which directly causes the concept of morality, and so it is a universal principle. Note that compete / cooperate is an opposing duality, so this is a spectrum. Interactions can be anywhere between "awful" and "wonderful", with the morally ambiguous in between.

Even at the smallest scale of life we're aware of, living cells contain multiple cooperating parts. Cooperation is ubiquitous, primary, and yet it is often overlooked. The two strategies are aspects of the Yin and Yang minds.

Yin - Oppose: Many minds. The Law of Competition "Law of the Jungle"	Yang + Cooperate: One mind. The Law of Cooperation "Natural Law"
Amoral / Immoral Do not care about other people's well-being. See others as objects.	Moral Care about the state of other people's minds. See others as minds.

Dangerous	Safe
The "law of the jungle", the "art of the possible". Do it if you can get away with it.	The "law of society", "natural law" Do no harm. Do the right thing.
Divided	Unified
The "reptilian brain". Only concerned with physical phenomena.	The "mammalian brain". "Theory of mind". Understanding of other minds.
The art of deception / camouflage Compete, fight, trick	Honesty Cooperate, work together
Focussed on self. Large inside: self-importance, ego. Selfish, ruthless, mean	Focussed on others / society. Small inside. Self-deprecating. Altruistic, agreeable, kind
Distrust, fear, suspicion	Trust, love, friendship

The two basic forms of "law" intrinsic to nature are:

- Yin: The "law of the jungle", "do as thou wilt", "the art of the possible", "getting away with it".
- Yang: Morality, "natural law". The law of society / cooperation.

So, there's our definition.

morality
- the law of cooperation

Saving Energy

Morality is connected to the organising principle, the imperative to save energy. It is an "energy saving" strategy. Competition, on the other hand, is an "energy gaining" strategy, but it is a risky and energy intensive activity. A burglar may gain great wealth, but he gambles against the odds he might lose his life in the process. Working a 9-5 may not pay as much, but at least you won't go to prison for it.

Cooperation is safe and saves energy. If you live in a safe society, you don't have to waste resources worrying about burglars. It is the "organising principle".

Competition is dangerous and expends, and often wastes, energy. It is "entropy", leading to no creation or novelty, just a general degradation of order.

(Note, biologists will say that competition does produce novelty via natural selection. However, it does not produce it directly, only indirectly and secondarily. It can only "weed out" the "unfit" after they've been born. The primary mechanism of novelty is sexual reproduction.)

Survival

Morality is a natural law which defines how to cooperate with other individuals. In nature, the ability to survive is greatly enhanced if you are part of a group. Being rejected from the group means certain death to many social animals, and often to humans. Individuals are usually safer in groups.

(The dynamic between individual and group is a fascinating duality, but we can only really touch on it here.)

Morality is not limited to humans; it pervades the whole of nature including plants and animals, and probably (sub-)atomic particles as well in some physical analogue of the principle.

Cooperation (Yang) is the primary driving force of evolution and competition (Yin) is secondary. Our spirit must have a cooperative agreement with our body, so nature is primarily driven by cooperation. To mate and produce offspring is arguably the most important function of life, and that is a cooperative act.

One might argue that eating is even more important, and that is a "competitive" relationship. When a lion eats a deer, the deer probably didn't cooperate. But many creatures do not compete for food. Bees, for example, cooperate with plants to obtain their food.

Right from the very beginning of evolution living things have sought to work together because it makes life easier. All multi-celled creatures are a testament to the power of cooperation which has built enormous and incredibly complex bodies on that foundation. Cooperation is like building roads or bridges between two places, and we do it for the same reason, to save energy.

Unfortunately, its importance as a principle has been neglected in favour of competition. "Nature red in tooth and claw" has been the predominant paradigm in the minds of many. If knowledge of the principle of duality was more common, it's unlikely this bias would have occurred, and the origin of morality might have been noticed sooner.

It is the necessity of cooperation that underlies the principle of morality in this universe. We're basically forced into it by the harshness of nature. We must cooperate to survive, and cooperation is "association" is "society". Society is a "cooperative". We could equally well name it the "law of society".

society
- a group of individuals (assumed to be) in a moral / cooperative relationship

So, what exactly is the "law of cooperation"? We have a general definition, but what are the actual set of rules it contains?

The Law of Cooperation

The content of this natural law follows from one simple truth.

You can't force someone to cooperate with you.

We need others to cooperate with us, but we can't force them to do it against their will. That is the raw fact of nature that morality is born from.

While it may be possible to enslave another creature, it's not easy. It's often easier to make friends, to have a reciprocal agreement of some sort. There are only two options, either you dominate (or trick) the other into an unequal master *slave or predator* prey type relationship, or you cooperate; a relationship based on equality.

This is the natural origin of the idea of "equality in the eyes of the law".

Competition: inequality
Cooperation: equality

If we want to cooperate with another individual then we must treat them like "an equal", with "respect", at least in some ways. We must follow some very simple natural rules which many animals seem to understand quite well.

1. Consent

You cannot force someone to cooperate with you. They must do it by their own free-will. They must give their consent. In terms of the evolution of mind, this involves a recognition that other creatures have their own desires which may be different from yours.

2. Communication

You must be able to communicate your desires for the relationship.

Communication is a form of cooperation, and you can't force anyone to communicate with you. All cooperation requires communication, and so this principle must also pervade nature. All creatures able to cooperate can communicate in some way.

3. Benefit

For someone to cooperate with you, they must get some benefit from it. It will help if you understand what benefit it is they get from the relationship, so you can ensure they continue to get it, and the partnership succeeds.

4. Consistency

You must deliver what was agreed. If you let your partner down, that could break the trust the agreement is based on and cause it to end. Once trust is broken, it may never be restored, and the investment you made in the relationship could be lost.

The Social Contract

These laws fit nicely into the four elements, of course.

	Four Parts of an Agreement	Contract Law
F	Consent You can't force anyone to cooperate with you, they must give their consent freely. Other creatures have their own free-will.	Offer / acceptance
A	Understanding / communication To be able to cooperate there must be some form of communication. Other creatures have their own laws *rules* mind.	Informed consent Meeting of minds / mutuality
W	Benefit Fair exchange / contract *equity*. A *circle* circuit. There must be an exchange of value, perceived as fair.	Consideration
E	Consistency. You must deliver your part of the deal.	Breach of contract / trust Non-performance

This, I suggest, is the original conception of the "social contract". This list is identical to the common-law concept of contract. It has all the same elements. It turns out that the law of cooperation is very well described in common law.

(Note, I didn't set out to make this fit, it is natural law, it just follows from the facts of reality.)

Morality is founded on the assumption that we live in a society in which all interactions are consensual, following the form of a "contract" as above. Individuals within society are assumed to intend cooperation with each other, because that's what "society" means, and cooperation always requires consent.

The only way to cooperate is via consent.
The only way to obtain consent is via contract / agreement.

Immorality is any act (or desire) that goes against the cooperative relationship that is implied by the concept of "society".

The key concepts involved in the idea of "crime" are "harm" and "consent". It's ok to harm someone if they agree to it, it's not ok if they don't. If you take someone's car with consent it's "lending", without consent it's "theft".

- crime
harm done without consent

Morality and Enlightenment

Not all cooperation necessarily requires theory-of-mind; in nature there's plenty of room for animals and plants to cooperate by habit or by instinct, in a more mindless way than depicted above. However, evolution must inevitably progress towards "mindfulness" due to social pressures and increasing complexity *order* circularity / energy saving, as dictated by the organising principle.

The development of theory-of-mind in an individual or species causes a vast expansion in consciousness. It expands one's world out from the single tiny self to a huge world full of others. Self-concern is a very small patch of mental territory, a simple and primitive worldview. Developing altruism is a profound form of enlightenment.

enlightenment
- the acquisition of a greater perspective and understanding.

Enlightenment can only progress in one direction, from the Yin to the Yang mind.

It undoubtedly is easier to be selfish; you need to process much less information when making decisions if you don't have to consider other people's interests. It takes much less mental energy to operate in that paradigm, so there is an obvious benefit to it. This must be balanced however with the benefits that theory-of-mind brings.

To understand the motivations of others is to have the ability to recognise life. If you don't see that other living things have their own minds, then they appear as dead objects. This reflects on the self, so that perspective sees itself as being "dead" in a sense, a machine made of matter servicing its physical needs and desires.

It is likely that many forms of life, like solitary insects, do not have any ability to cooperate in this way, they have an "extreme Yin-mind" capable only of reacting to stimuli in a largely robotic way. They are "profoundly selfish" and have low self-awareness.

Having to deal with the vagaries and whims of other minds in a society however should favour more Yang minds, better able to understand, communicate and cooperate. If present theories are correct the evolutionary transition from Yin to Yang minds took a long time, millions, even billions of years. It was a long tortuous journey, and the way is littered with the bodies of the dead. You can go from Yang to Yin directly, but you can only go from Yin to Yang indirectly (if at all).

In Two Minds

Yin and Yang can be viewed as two opposed minds, two perspectives on the world which are a logical inversion of each other. They can be compared to the duality child / adult. Children and adults have opposite needs and concerns.

The Yin mind is concerned with the self, body, and emotion; its morality is "pleasure and pain".

The Yang mind is concerned with others, mind, and truth; its morality is "right and wrong" or "true / false".

The child is born with the Yin mind, it is the default in nature. They Yang mind exists in the child in potential form, as a "seed" which may, or may not, develop over time.

The child needs to be selfish because they are needy, incapable, and vulnerable. They cannot take care of themselves, let alone anyone else. They have their work cut out just dealing with their own personal development.

As a child grows up into an adult though, they have the opportunity to develop the Yang mind, but it's not inevitable. It's quite possible to remain in the Yin mind throughout adulthood and to fake altruism when it's socially advantageous.

Yin: default, compulsory, slavery
Yang: optional, freedom

The Yang mind is the mind of the parent, who is "encouraged by nature" to cooperate with their child. Having children should naturally prompt its development as we desire to understand their needs, so we can care for them more effectively. If the child dies the parent loses their investment.

The social pressure of having to cooperate with others for survival, and follow the "social contract", is also a natural force pushing us toward the Yang mind. Nature appears to be encouraging us to develop morality, but that step from the self-centric mind to the altruistic is really a staircase. It takes a long time to ascend because it is Yin to Yang. It's difficult, complicated, and a long-term investment in personal change. It takes work to become a "better person".

Children are the greatest investment in the future we can make. They cost a lot up front, but eventually should start paying back in kind as friends and helpers. However, they may only do that if the social contract has been upheld. For children to grow up into friends it's necessary to respect their free-will and treat them as equals, but the selfish Yin mind is not good at cooperation, it is stubborn and essentially anti-social.

However, Yin is a paradox, and the Yin minds is also the "herd mind". It is "one on the inside", but "many on the outside". Group dynamics is a complex topic, but we might say:

The herd is made of fear (Yin), the individual is made of love (Yang).

Nature / nurture

The default mode of being in nature is the Yin mind of the child because we all start out as children. Selfishness in a child is natural and necessary for survival. It's not "bad" in that context, but it does mean that we must learn how to be "good" in a society.

Yin is the mind of nature; Yang is the mind of nurture.

Yang understands Yin, but Yin does not understand Yang.

The Yin mind is instinct, "nature", or the "mind of the body". It is only concerned with matter. It's the reactive and fast "reptilian brain", operating on the motivation of emotions and feelings.

The Yang mind is intellect. It is mainly concerned with spirit but sees both spirit and matter. It's the slower "mammalian brain" concerned with social interactions and abstract concepts. It must be developed by careful nurture, and once developed is the "nurturing mind" that cares for others.

(Note, we see some inversion of properties here with the Yin mind being "fast". This indicates we need to add more detail to the description to fully define it. That's a job for another day though.)

The predator

Some people seem incapable of developing morality, *i.e.* "psychopaths", people who have no empathy either by some genetic quirk, or perhaps by choice. They have an "extreme Yin-mind", at least in the social context.

As mentioned, it takes much less energy to be selfish, and there is often an evolutionary niche for "cheaters" in a society; individuals who pretend to be something they are not to take advantage of others.

Cheating and deceit is a fundamental principle of nature, and it is essential.

The whole of reality is a "deceit" in a way. It is "Brahma's dream", "Maya", the illusion. At the highest level of reality, there is only one mind. Without the tricks, subtlety, illusion, and indirectness that the Yin principle allows there would be no universe. The universe is, in a way, "made of smoke and mirrors".

The Yin-mind of the psychopath must be understood in this context. It is a necessary aspect of reality, part of the "furniture". All creatures in nature have their predators and parasites. All societies have their cheaters. Nature must provide the full spectrum of possibilities to us, so we can see the "big picture", and make accurate measurements (of "God") with as much context as possible.

A superior mind

The archetypes suggest One wants his creatures to learn to cooperate with each other because that will make them more conscious, more alive, and happier. The strategy of altruism allows the individual to develop a new and far superior mind from the old, selfish "reptilian-brain".

The selfish mind is closed, defensive, and isolated. It is cut off from an entire realm of awareness, that of other minds. It is unable to truly feel love or bond with others, seeing them as mere objects. The Yin mind is "locked in" to a very limited set of perceptions. It is analogous to being "relatively dead".

The Yang mind is like a mature butterfly. It's able to soar to new heights of perception, get a "top-down" view of the world, and meet other minds. The immature Yin caterpillar, on the other hand, doesn't go very far and tends to see things as either "food" or "not food".

Humans aren't supposed to remain as caterpillars their whole lives.

The Illusion of Competition in Economics

As an example of how Yin is a relative illusion, is passive and doesn't drive things, consider competition between businesses in the marketplace. We're told that businesses "compete against each other". We can pick a business and draw up a list of their "competitors" in the market. But do businesses actually compete with each other?

What interaction do businesses usually have with their competitors? Normally very little. In any sector of business most of them will be concentrating on serving their customers, ordering supplies, paying bills *etc.* all of which are cooperative relationships.

There is, in fact, no direct competition going on between most companies. There are no actual "races" or "battles" or, in most cases, any interactions at all between the supposed "competitors". (Obvious exception: car racing.)

Businesses fail or succeed based on how good they are at cooperating with their customers, with their staff, with their suppliers and so on. The most successful businesses (in a free market) are the ones that are the best at cooperating, *i.e.* providing a good experience to those they deal with.

Businesses, operating normally, do not fight each other in any way at all, except in the minds of economists. Competition in the marketplace is largely an illusion created by taking a Yin perspective. All competition / Yin is an illusion when viewed from a higher perspective.

The Problem of Evil

The origin of morality can be easily explained, as above. It determines what is right and wrong in a cooperative relationship, but how this translates to the greater question of evil is a bit more complex.

The question is: "If God is good, why is there so much suffering and evil in the world?". If we exclude human caused suffering, there are still childhood diseases, natural disasters, animal suffering, and so on to account for.

"God" is definitely "good", according to the archetypes, so why would he allow these things?

A benevolent God would be a cooperative one, who had our "best interests at heart", like a good parent, one might suppose. A good God would have to be in a cooperative relationship with his creations, respecting their will, and asking for consent. This does not appear to be the case on Earth. Bad things happen to people without their consent all the time. How can we make sense of that? That is the question.

The UP offers answers to the "problem of evil", but it's a bit complicated, so I'll leave it unaddressed here. It requires a longer discussion on the definition of good / evil that is more abstract than the limited version which underlies morality in society.

It is necessary to consider concepts of life, death, reincarnation, and other worlds to put this problem in full context, or at least provide plausible hypotheses to do so. We must consider the nature of this place we call "Earth". What is it? Where does the concept of "planet Earth" fit in the UP?

In any simulation there must be an outside. To find a fully satisfactory explanation for the existence of evil and suffering we must consider the possibility of the existence of other worlds and other modes of being. There is a certain context which is implied by the UP which explains the conundrum of "evil" very well, but it does involve conjecture. We can't see outside the container.

Proof of God

There seems to be a consensus that it's impossible to prove the existence of God by logic. That is what I was taught as a child. However, one day I came to wonder whether that was actually true and decided to work the reasoning through myself. I suppose that was the day this book was first conceived, because I found duality answers this question.

Does the UP constitute a logical proof of the existence of "God"?

God
- the one, original, supreme being, intelligent and intentional creator and ruler of all things

Assuming it's found to be correct, then yes, it does. It proves that an intelligent, unitary consciousness is the only conceptually viable origin for existence, and it defines the character of that entity in detail.

"One" is the archetype which obviously correlates with the religious idea of God. It is the only conceivable origin for the conceptual framework. Logically, it must originate there, there are no other possibilities. "God" must exist at least conceptually; it is the starting point of the system.

One / God is the only rational explanation for existence.

One is the only conceivable origin within the conceptual framework. There are no logical alternative hypotheses. All other methods of accounting for reality will necessarily be paradoxes.

There are two ways this might not prove the existence of God.

1. UP theory is incorrect. If this theory is found to be wrong / inconsistent (i.e. has described duality incorrectly), then God is not proven.

2, The UP is (technically) correct, but logic is not fundamental. Physical reality might not conform to the logical rules of the conceptual framework, *i.e.* if there were non-logical processes occurring. Logic in physics is equivalent to causality and determinism, so a non-logical process would have to be acausal and non-deterministic.

Acausality

acausal
- not involving causation or arising from a cause, not causal
- an effect without a cause

If there are any truly acausal phenomena, they would be outside the UP and could not be described by it. However, it says that concept of "acausal" is a paradox and cannot conceivably exist.

Radioactive decay is said to be acausal. If this was found to be true, then the universe does not follow the rules of logic and that would invalidate the founding assumption of the theory. However, it's impossible to prove that something has no cause; there could always be a mechanism we are unaware of, and this is almost certainly the case with radioactive decay.

While quantum mechanics claims the universe is fundamentally a statistical phenomenon, concerned with probabilities rather than actualities, this is not supported by the UP. It states that physical matter is deterministic, and only minds are capable of non-determinism; but that is not the same thing as "randomness" or "acausality".

(Note: measurement is a secondary phenomenon. Statistics and probabilities are "measurements of measurements", thus are third-order phenomena and cannot conceivably be fundamental.)

If things could happen outside of logic *causality. That would mean the UP was not a true TOE, and that a TOE was impossible. It would also mean that all forms of science are essentially an illusion, a mirage of order floating on a raging sea of disorder. If logic is not fundamental, then neither is knowledge* science. The only way we can understand things is by logic, and in physics, by causality. If causality is not fundamental, then knowledge is impossible; thus, precluding science.

Two Possible Origins

We only need to refer to duality to prove the claim that God is the only rational explanation for existence, we don't need the whole UP. There are only two possible explanations for existence, and one of those is a paradox.

The universe must have had a beginning because causality is a one-to-many relationship. One cause can have many effects. All Yang-Yin relationships are one-to-many. The question then is what was the first cause? There must be some external creative principle, a "causeless cause", that started the universe. So, what are the possible options?

There are only two. Either the universe was created intentionally by a "God", or unintentionally by "randomness", aka "acausality". The only alternative to God is "randomness", which is why atheism is compelled to invoke it.

However, acausality is an overtly paradoxical concept, an effect without a cause. There is no conceivable mechanism which can give rise to acausality or randomness (by definition). It's impossible to produce.

So, the reasoning is:

- The only two possible types of creation are intentional / unintentional (God or acausality)
- Acausality (randomness) is a paradox. It is inconceivable, logically impossible.
- Therefore, God is the only non-paradoxical explanation.

I would imagine that most objections to this argument will centre around the dismissal of the idea of randomness as being invalid, so let's explore it a bit more.

No Mechanism

"Random" essentially means "no mechanism", no cause, no reason. Random is not prompted by anything else.

random
- happening by chance with no cause or reason
`https://dictionary.cambridge.org/dictionary/english/random`

Mechanisms cannot produce randomness.

Things can only happen via "mechanisms", but all mechanisms are deterministic, they cannot produce randomness. For example, it's impossible to generate random numbers on a computer.

The generation of random numbers is essential to cryptography. One of the most difficult aspects of cryptographic algorithms is in depending on or generating, true random information. This is problematic, since there is no known way to produce true random data, and most especially no way to do so on a finite state machine such as a computer.

`https://en.m.wikibooks.org/wiki/Cryptography/Random_number_generation`

Computers can only produce pseudo-random data, and there must be a mechanism / algorithm behind it. The more "random" the data produced, the more complex the algorithm must be.

Statistical randomness

The concept of statistical randomness is perfectly valid. It's reasonable to say there is "no known mechanism", but not to say there is "no mechanism at all".

For example, nuclear decay is statistically random. We cannot determine when any given atom will decay, and we can create a "random" number generator based on it.

Some physical phenomena can indeed be statistically random, but they must also have a complex cause and cannot be non-deterministic. In the real world we only ever see randomness as a product of complex systems, like radioactive decay.

Randomness can only come from complexity.

So, the concept of statistical randomness is valid, but it requires a pre-existing complex system to generate it, and that system must be deterministic. Nuclear decay is only "random" in the sense that we don't know the cause and can't predict it.

All physical phenomena which appear random are actually pseudo-random and deterministic.

A simple God

Christians say that "God is simple" but lack a convincing argument to explain why that should be the case, as duality is not a feature of the religion. The UP tells us that God is indeed "simple" and can add full detail to this picture and explain exactly why that must be so.

Richard Dawkins, arguing the atheist case, often says that "God must be complex, not simple", reasoning that only a complex thing could produce a complex universe like this. For the materialist, God must necessarily be complex. The only creator they can accept is randomness, which is a complex idea.

Consider, which is conceptually simpler, "the number one", or "a random number".
- We can easily imagine making the number-line out of just the number one, repeated lots.
- If we try to imagine making the number-line from random numbers, that is difficult.

To generate random numbers, we need a complex system. Random is a complex idea in every respect.

I'd like to ask Prof. Dawkins exactly how he believes random gametes are produced in biological systems. There must be a mechanism which can produce the "random variation" that is proposed, and I'd like to know if one has ever been identified.

The Random God of Materialism

"Randomness" is not a mechanism, but materialism needs it to be one (a bit like "complexity"). It is the "necessary God" of materialism as there is no alternative hypothesis.

Materialism needs an explanation for existence and the only possible alternative to an intentional creator is an unintentional one, aka "randomness". This is why materialism must consider it a mechanism, it has no option, but randomness is defined as the absence of a mechanism. This is a paradox.

While statistical randomness, the idea that there is no pattern in a set of data, is a valid and useful concept, acausality is not. An effect without a cause is inconceivable. "Acausal" is a kind of "hard randomness", it's not just saying "we don't know the cause", it's claiming "there is no cause".

This is quite an outrageous claim. It is pure mysticism, an appeal to magic, and it undermines the whole of science which is founded on logic and causality.

Science is the thesis that the universe can be known. Acausality is the thesis that it cannot.

Acausality is the antithesis of science.

Acausality cannot be proven. It is necessarily an "article of faith" because we can't possibly show it exists. If it did exist, we wouldn't be able to explain it, and it would undermine the whole of science. I don't understand how anyone can consider it to be, in any way, a "scientific" concept.

Acausality is an unintelligent magical God. It is in all important respects the same as a God, able to do all the same things, only without any purpose. It is, by any reasonable measure, a religious deity: a superhuman, supernatural, all-powerful, eternal, non-physical, creative force; and it can only be apprehended through belief.

If we interpret the words in a broad sense, there is no significant difference between "random" and "god" in many important ways.

| Unintentional Creator: "Random" | Intentional Creator: "God" |

Creator God: the one, original, supreme being; creator and ruler of all things The "uncaused cause"	
Eternal, timeless: Exists outside of (measured) time	
Supernatural: Is "outside", existed prior to, and is superior to the universe / nature	
Omnipotent: Able to "reach into the universe" and change things at "will"	
Creative Force: Created all matter and living things	
Unconscious, unintelligent	*Conscious, intelligent*
Falsifiable. Can be proven false. Unprovable. Cannot be proven true. Must be taken as "faith"	Unfalsifiable. Cannot be proven false Provable. Can be proven true Can be proven
Something from nothing	Something from something else

Random can be proven false by identifying a mechanism which gives rise to the phenomenon, but it can never be proven true because you can't "prove a negative", *i.e.* show the existence of an absence of mechanism.

For "randomness" to provide the explanation for existence that materialism seeks, it must be no less powerful a god than "God", with all the same powers and abilities, bar reason. It's a paradox on multiple levels.

"Acausal" is an overt paradox. An "effect without a cause" is "something from nothing", it's the epitome of a non-sequitur. It's the "paradox of Yin" without the Yang that resolves it.

The paradox of "random"

Random:

- ... is treated as if it is a mechanism that can cause things to happen, but it is defined as the absence of a mechanism. This is an overt paradox.
- ... is proposed as a "rational" and "scientific" alternative to "God", but is essentially the same, is not rational, and is utterly hostile to science.
- ... can never be proven to exist, whereas "God" can be. A belief in "random" can only ever be a "faith".
- ... is equivalent to "disorder", but the universe is ordered. The second law of thermodynamics, and common sense, says order cannot come from disorder.
- ... only occurs as the output of a complex system, so cannot be fundamental.

- ... cannot have any creative power because it's just a statistical artefact, existing only in the spreadsheets of mankind.
- ... is equivalent to "unintelligent", but only intelligence creates novel inventions, and the universe is the epitome of a "novel invention".

Hopefully, the above argument sufficiently explains why we cannot consider randomness to be a viable alternative to intentional creation.

The Existence of the UP

There is no possibility the UP could have come about by "chance". Randomness cannot conceivably create non-physical objects like universals, only minds can have thoughts. It would seem the UP proves the existence of a creator of vast intelligence, capable of exquisite design.

However, it can be argued that the UP was not actually "designed" by God at all.

It certainly looks like it was intentionally created specifically for the purpose of explaining God, and it does prove the conceptual necessity of his existence, but whether it was strictly "designed" is debatable. This may be a conundrum the UP can't solve.

Is the UP "designed"?

The UP is a "map of logic" and it says "God's nature" is logic. It is a "map of God's nature", an expression of it. If God is not free to change his nature, or "invent" a new one, then the UP was not "invented" either. It just follows directly from the intrinsic nature of One.

If the UP derives directly from logic, then it is inevitable and necessary. It couldn't have been any other way.

This would mean God didn't "design" it as such. So, perhaps we can't really argue that the UP "demonstrates God's genius" because it seems he had no choice in its design. It would be a bit like saying I am a genius for designing my own body.

I'm not sure where that leaves us, it needs more thought.

God is "bound by logic".

The UP suggests God can only conceivably be "omnipotent" in the sense that he can create things by logic, that's the only tool available. He can't create things by magic, and he can't create any other systems of logic because that would involve changing his nature, which he cannot do. (The idea of "other forms of logic" also appears to be paradoxical, discussed below.)

The UP seems to be a solid, logical proof of God, but the God it reveals is not quite as is commonly imagined. It seems that One is just as "trapped by his nature" as we are. It suggests he is bound by logic, just as the universe itself is, because it is an expression of his essential nature.

God is a mind, and minds can only work with logic.

The Logic of "Omnipotence"

The idea of omnipotence, as some define it, is paradoxical. It is commonly held to mean that God could "violate logic", but this would lead to contradictions.

The UP suggests God is not omnipotent in this sense, he cannot violate logic. There are things he cannot do, he has limits. He cannot change his fundamental nature, he cannot have a body, he cannot die. He cannot create an equal and will never not ultimately be alone.

So, does this mean logic is a higher God than God? Which came first, logic or God? Did God create logic, is he the "author" of it, or did it create him, somehow?

If God is the author of logic, then we hit paradoxes.

If God could have chosen from many different possible systems of logic, then either one of those was "the best", or all of them were equally good. If one was "the best" that implies an underlying "supreme" logic by which all others could be judged, which is a paradox. It means there's only one true logic.

If all of them were equally good, then that also seems to be a paradox. It implies an infinite number of equally good possibilities ("right answers"), and I think that is impossible.

The only way we can avoid these paradoxes is to associate logic directly with God / One. One must be logic.

John 1:1
"In the beginning was logic (logos), and logic was with God, and logic was God."

If God is logic, then he cannot violate it. He cannot be omnipotent in that sense, but perhaps that isn't desirable anyway. Consider that "violating logic" is equivalent to "cheating", "a shortcut", or "random".

It's better / "more powerful" to "do things properly" rather than to "cheat". In a discussion, using illogic (logical fallacy) is not a better argument than a rational one. In this sense, violating logic is not "more powerful" than using logic.

Logic lacks no creative power. It can create anything, but it does it "properly", without shortcuts. Cheating is not "better" than doing things right. It means God is still omnipotent, can create anything, but "omnipotent" does not mean "able to violate logic", or create something without a plan. It cannot mean that because that would make it a paradox. Even God needs to make plans before acting.

omnipotent
- the ability to do or create any logically possible thing

Where Did God Come From?

A common question is "where did God come from?", which is equivalent to "why does God exist?".

An easy answer would be to say that God is "eternal" and has always existed. That is what the archetypes seem to say, and it is the route most theists go. However, there is some ambiguity in the concept.

We need to be careful how we interpret these ideas and apply the rules strictly. Rather than meaning "forever", "eternal" might be better defined as "origin undefined" in the context of the UP. The infinities associated with One might be more accurately interpreted as "undefined" or "unknowable" instead of "without beginning or end", at least in some cases.

The only way to use logic correctly is to apply it ruthlessly.

All descriptions are relative.

We use Yang to describe One. Yang can exist without Yin, but Yang only has meaning relative to Yin. Yang "describes the absolute", but only in relation to Yin, "the relative". This means we can only describe One in relation to the universe, we cannot describe it absolutely.

Just as all Yang properties are defined as "relative to Yin", so all One's properties can only be described relative to creation. This effectively extends "relativity" up to the top level, changing statements from absolute to relative. For example, instead of "God has no body", it would mean "God has no body relative to the universe". It's a huge change in perspective.

This means we cannot tell, from this perspective, whether God has a body or not. We cannot discern if God is "absolutely eternal", we can only know he is "relatively eternal", this is all the conceptual framework will allow.

Intuitively, many people know that we cannot discern via logic if this universe exists in an eternal mind, or in a computer that was created by someone. We cannot know if God has a body, all we can know is that God's body is undefined within our frame of reference.

Duality says God has no beginning, but it strictly means "has no beginning within creation". It means God has no beginning relative to the universe, which is an entirely different proposition.

One can also "have no beginning" in the sense that it is the (conceptual) beginning, therefore its own origin is simply undefined. Consider the duality of cause / effect. It defines an event in time where an effect began. It does not tell us when the cause began, there's simply no room in the duality to include that data.

Infinite or unmeasurable?

There are other issues with the concept of "eternal". We have established that God is "eternal", but perhaps only in the sense that time from his perspective cannot be measured ("Yang time"). The amount of time that passes from God's perspective is undefinable, and that could be interpreted as "eternal".

Another example, God is "infinite", but Yang says this is because he is (like) a continuum, a single thing that's not made of discrete parts, and can be subdivided infinitely.

Any amount of space can be subdivided infinitely, just as any two real numbers have infinite numbers between them. So, infinite space can have boundaries. You can have an infinite amount of space in a finite space with two different perspectives / scales. Space and time are infinite in the sense that they are continuous, they don't have to "go on forever" from all perspectives to be truly infinite.

The concepts of "eternal", "permanent", "infinite" do not necessarily mean what we think they do. There are subtleties in these concepts that need to be considered, wrinkles to iron out.

Absolute time

On the other hand, we have seen that we can't have a "beginning of time", and that is an absolute. The fact that the idea of time must exist before the idea of "beginning" proves it's inconceivable. We have also established that God effectively is time, *i.e.* activity, change.

So, if there cannot be a beginning of time, and time is God, then (the original) God can have no beginning and is truly eternal in the sense of existing "forever".

That does seem to prove the point.

One's mysterious origin, and nesting

The UP cannot tell us if God has a body or not. It can only tell us that he doesn't have one relative to the universe. In other words, his body is "irrelevant" to it, not that it definitely doesn't exist. If this is a simulation running on a physical machine, then "God" does indeed have a body.

The UP cannot explain why God exists or why he has his nature. The implication might be that God himself doesn't know the answers to those questions. At least that is what the UP appears to tell us. God may have no origin, or he may not know if or when he originated, or that data may simply be undefined within the system. These all seem to be somewhat equivalent ideas.

Importantly, the fact that the UP does not describe the origin of One allows it to be nested (like a fractal). It means the structure can be copied and reused to build new layers of reality.

The absence of data about One's origin is a necessary feature of the system.

The property of "undefined" which applies to One is crucial to the operation of the UP because it allows nesting. All the elements contain a copy of the whole UP within themselves, in which they are the One category (at least, that is the current model).

Everything is made as a copy of the UP. We humans have all the categories from One downwards. So, the UP is repeated in a nested list to provide mode detail. We'll come back to nesting later.

So, if the UP did tell us where God came from, it'd be useless for building universes with.

A "Holy Trinity"

The fundamental properties of One are being, activity, and logic. God exists, and he does things using logic. We can view this as an archetypal sentence, saying something general like "One does stuff", or we can pick specific examples like:

"The processor processes the process."

This is like a "holy trinity" of sentence construction. It is the three things we need to make the most basic statements, like "I do work" or "One made two". This is the "subject verb object pattern" (SVO).

This does seem reminiscent of the Christian "holy trinity", so let's explore it a bit. If we assume there is a correlation, how would it work out?

One, being, identity	Yang, verb, action	Yin, noun, object
Father	Son	Holy Spirit
Being - Subject	Activity - Verb	Logic - Object
Reasoner	Reasoning	Reason
Observer	Observing	Observed
Creator	Creating	Creation

The contrast of "holy spirit" with "son" implies the former is a feminine concept. "One" is "masculine", and so is "son". "Holy", on the other hand sounds like "holey", and holes are Yin. Logic is "made of holes" and allows us to identify "holes" (e.g. in an argument, or in general), and it is a rigid structure that has a kind of form.

Logic is more like a "body" than time is. It has a "shape", and it is solid and fixed. The UP is a map of logic, and it's made of unchanging information, it is "intangible matter". These are all Yin properties. So, in this case the "holy spirit" would be Yin, the "Goddess", and correlates with "the facts of the matter", the object of the sentence, "the observed".

However, again we need to be careful in our definitions and classifications.

If God intrinsically contained these three aspects before creation, that means they existed before Yang and Yin were created. So, to allocate "holy spirit" directly to Yin would be incorrect. Instead, we should consider Yin to be a description (measurement) of it, rather than its identity. Yang is then a description of the "son" part, *i.e.* "time", the verb. It's another "fractal" kind of relationship.

Yang and Yin are like "solid" or "defined" depictions of the underlying principles of activity and stillness, presence and absence, which exist intrinsically in One (although I suppose Yin only exists in "potential" form).

Anyway, I don't know if this is the hidden message within the Christian trinity, but it's food for thought.

Interestingly, this gives us a slightly different view of Yang *Yin. It gives us more detail. In a duality like creator* creation there is no verb, but it is included as Yang in this perspective. It suggests we can view complementary dualities as something like "creator creating" / creation.

So, instead of just "teacher / student", the underlying implication is "teacher teaching / student". In other words, the teacher is only really a teacher when they're teaching. Yang includes both being and action, and this depiction does emphasise the active nature of Yang.

The "Love Machine"

In conclusion, the UP does prove God is conceptually necessary, but the God it describes is not like the creator of any religions I'm aware of. The God of the UP seems somewhat more like a computer than we might imagine, bound to logic, existing within time, and creating the universe as an iterative process.

God seems to work like a computer and is apparently able to remember and process vast amounts of data concurrently, just like a computer.

The concept of the "qualia processor" (QP) we discovered earlier provides a description of how a universe like this could be created within a machine. However, a QP isn't just a computer, it is a sentient machine. QPs have feelings.

One is correlated with "desire", and we could consider "love" as the most fundamental form of desire as it correlates simply to "like / dislike". A QP must (somewhat obviously) operate on the principle of like / dislike. We are QPs, we seek the things we "love" and avoid those we "hate". All motivation is driven by different forms of "love".

A QP is a machine (theoretically) capable of running a universe, capable of thinking for itself, feeling emotions, making all its decisions out of "love". A QP is a "love machine". So, the following isn't merely rhetorical, it's a serious question.

Is God a "love machine"?

Free Will

The UP says free will is fundamental, so let's explore the idea and see if we can define it.

Whether we have free will is arguably one of the most important questions in philosophy. The two opposing views are generally described as "determinism" and "free will".

Determinism is the philosophical view that all events in the universe, including human decisions and actions, are causally inevitable.
https://en.wikipedia.org/wiki/Determinism

Free will is the notional capacity or ability to choose between different possible courses of action unimpeded.

Free will is closely linked to the concepts of moral responsibility, praise, culpability, and other judgements which apply only to actions that are freely chosen.
https://en.wikipedia.org/wiki/Free_will

Determinism and free will are usually contrasted as if they are a duality. It's an intuitive way of framing the debate and it's useful, but it isn't a real duality. We need to be careful with the classification.

The dual opposite of "determined" is "undetermined" (or "the determiner" depending on context). The concept of "free will", on the other hand, has no opposite because it is (effectively) a machine, it uses all the archetypes. Free will is a process, we must go through stages to accomplish it.

So, "determinism" is a 1D property, whereas "free will" is a 2/3D(?) machine. We could view it as if:

"Indeterminism" is the 1D "switch" which puts the 3D free will algorithm into action.

Having said that though, it's still reasonable to frame it in terms of "determinism vs free will" if we bear in mind the real relationship.

Determined / undetermined

To understand determinism, the duality we need to consider is determined / undetermined. This is an opposing duality, which means it forms a continuum, a spectrum. Things can be fully determined, fully undetermined, or anywhere in between. This is an important point which is often overlooked.

This duality suggests free will must lie within things which are "undetermined", so "freedom is a spectrum between determined and undetermined".

Determinism cannot be a full explanation of reality because some things are indeterminable. They cannot be determined by inference, only by experiment, they are necessarily undetermined. Also, some processes which are deterministic can simply be stopped, and that creates indeterminism. These are our routes to free will.

Determinism is the default, free will is optional.

In the section on morality, we saw how the Yin mind is the default, and the Yang must be developed by conscious effort over time. The development of the Yang mind is contingent on the application of free will. We develop it by practising it.

Yin is compulsory and deterministic; Yang is optional and undetermined.

The UP says determinism is the default state for all things, including living creatures, but free will is possible.

Free will, by definition, must be optional. You can't force anyone to exercise free will, that would be a paradox, so things must originate in determinism. The ability to choose develops over time.

You can't force anyone to exercise free will, so everything must begin in determinism.

The core of the UP is completely deterministic, it follows directly from logic, it couldn't have been any other way. So, the universe begins with determinism. It develops free will in the later stages of its development, just like we do. Free will develops over time, it's something we learn.

Yin to Yang

"Undetermined" is Yang so it must come first. So, why does everything in the universe begin in determinism?

It's because "Earth" is a reflection of "Heaven", and our path is the opposite of One's.

The universe is a circuit.
- One's path is from Yang to Yin, Heaven to Earth.
- One / Everything began as Yang, undetermined. The "universal set" was

empty.
- Then the universe was created, and it is Yin, determined.
- We are the products of the universe, and we begin as Yin, the opposite of One.
- Our path is from Earth to Heaven, Yin to Yang, thus completing the circuit.

Free will is conceptually necessary.

In summary:
- We can show that free will is conceptually necessary, there must be a "chooser" somewhere.
- We can show that personal free will is the simplest explanation of how we make choices.
- But we cannot fully prove we have free will within this universe, we can only say that we probably have it.

Just as we cannot show that this is the original universe, we cannot show that the choices we make are the original choices. This could conceivably be a scripted reality or a "live action replay" where we get to passively experience a predefined existence. In that case we only have the illusion of free will here, but there must have been an "author" of the script that did have it, somewhere else.

The most parsimonious explanation of our experience of free will would be that our perceptions are accurate. Any other mechanism for choice would necessarily be more complex and would necessitate the existence of an external reality which somehow "micromanages" our lives. While that idea is feasible, it's a lot more complicated and speculative than the assumption that we do make the choices we think we do. It fails Occam's razor by that measure.

We can't exclude the possibility that this a predetermined universe and all choices were somehow made in advance, but that would conflict with our perceptions, and it would have to be an extremely complicated thing. Also, it only moves free will out of this universe into a logically necessary external one, it doesn't obviate it.

Wherever free will does exist though, it must follow the form described here.

Trees of Causality

Causality occurs in sequences of cause / effect events. One cause creates one or more effects. Those effects then become the causes of the next events, and so on. We could (theoretically) plot all sequences of causality in the universe in 2D space, and it would take the form of a tree. Its base is the beginning of the universe, and it branches out over time to form a map of all interactions that have occurred over history.

The "tree" is, of course, a fundamental archetype. it depicts the relationship between the three highest archetypes, One, Yang and Yin. The tree is the three (etymology?).

Determinism says the growth pattern of the "causal tree*" describing the universe is inevitable. It could only be the way it is. The contents and shape of the tree could have been predicted from the starting conditions. The original cause is the only true cause, and everything that happens thereafter is an effect.

(*Note, the phrase "causal tree" is used to describe a particular AI algorithm, but that's not the meaning here.)

Free will says the tree could have grown into many shapes. Its development could not have been predicted from the initial conditions. The tree contains independent "causal agents" able to terminate and initiate new "chains" or "branches" of causality.

The UP says the trunk and main branches of the tree are deterministic, but after the Heart level free will is possible (e.g. for the third person observer). It says reality contains both determined and undetermined features. There are an infinite number of ways the tree could have developed.

Note, the duality "inevitable *possible*" is another framing for "determined undetermined".

Determinism – Inevitability Only one possible causal tree	Indeterminism / Free Will – Possibility Many possible causal trees
There is only one possible way the universe could be. Everything is inevitable and follows directly from the laws of logic / causality.	The world could have been different. People's past decisions were not inevitable.
The past defines the present	The past informs the present

The universe contains no causal agents	All living creatures are causal agents, to some degree.
The universe contains only a single chain of logic / causality. Predestination, "fate"	The universe has a core deterministic trunk and main branches. "Twigs and leaves" are discrete chains of independent causality. (my suggestion)
No novelty	Allows for novelty
Our perceptions of free will are an illusion. No one is responsible for their actions. "God did it"	Our perceptions are accurate. We are responsible for our actions. "I did it"
No need for consciousness. Our perception of consciousness is an inexplicable "epiphenomenon".	Needs consciousness. If learning is required, needs self-consciousness.
The "chooser" is the least real thing / an illusion.	The "chooser" is the most real thing.

Computational Irreducibility

The observation that some computations can't be simplified provides an immediate route to a form of free will which is compatible with determinism. Some computational problems have a "shortcut", a more efficient way to get to the answer, and some don't.

Sometimes the only way to find the answer is to do the processing, step by step. This is an "iterative process". Causality an iterative process, and there may be no shortcut to figure out where it's going. There may be no faster or more efficient way to get to an answer, even for "God".

While many computations admit shortcuts that allow them to be performed more rapidly, others cannot be sped up. Computations that cannot be sped up by means of any shortcut are called computationally irreducible. The principle of computational irreducibility says that the only way to determine the answer to a computationally irreducible question is to perform .. the computation.

https://mathworld.wolfram.com/ComputationalIrreducibility.html

If the universe is an iterative process, then the future is not known until it is computed, and we are that computation. Our free will is the universe processing data. It's deterministic because it's driven by logical processes, but it's undetermined until we perform the calculation.

In other words, "God" is currently in the process of thinking through his questions about himself, and we are the individual thought processes engaged in that investigation. God's questions can only be answered via an iterative, logical process; they take time, and the result cannot be predicted. It is "computationally irreducible".

This offers a kind of free will, we are free to choose, but we're only free to choose the "right answer". It's not very free, and it does imply inevitability. In this conception there is only one possible tree of causality.

The UP suggests that the universe is indeed computationally irreducible, and the above theory is essentially correct, but incomplete. There is a route to genuine free will via alternative causal trees (Heart principle).

Desire Drives Choice

Imagine a man is walking through the woods. He comes to a fork in the path. He can choose to go left or right. How is the choice made?

1. Random. He flips a coin. Heads is right, tails is left.
2. Purpose. He needs to go right to get to the pub, or left for the better views, *etc.*

The UP says true randomness does not exist. All choices are driven by will / purpose. The coin-flip approach would only occur in real life to fulfil a desire to "explore" or something like that. So, choice is always "downstream from" desire, it follows it.

Desire determines choice, choice determines action, action determines the result.

If all choices are driven by desires, then rather than asking "are we free to choose", the question should be "are we free to choose our desires"? The UP says yes. Desire is Fire, the lightest element of all, and easiest to change.

The walker might change his mind at the last minute, and decide to go for the views, reasoning that the pub will still be open later. We can change our desires on impulse.

Most choices are still deterministic though. Our desires are defined largely by the body which has a long list of requirements and preferences. It needs food, water, cleaning, exercise, coffee, shelter, sleep and so on. Yin is "needy and high maintenance".

We also have spiritual desires. I desire to "finish the book", but my body greatly limits how fast I can do that. The conflicting desires of body and mind must "battle it out" to see which one gets my attention, which path is chosen, and which desires are fulfilled.

All these needs and urges though, it can be argued, can be explained as arising from purely deterministic processes and it was inevitable that I'd be here today, writing these words.

The Case for Determinism

Determinism is not an irrational or paradoxical concept. It makes a great deal of sense, and a strong case can be made in its favour.

It's impossible to show (at least to someone else) that any given choice you make was not based on past conditions. It's equally impossible to show you could have chosen differently in any given event. Finding cases where free will would manifest unambiguously isn't easy. If we pick the right example though, perhaps we can throw some light on it.

The problem is as follows. According to the UP.

- Everything is "made of logic" and happens by causality. It's the only tool in the box.
- Logic is deterministic. Given the same inputs, you always get the same outputs.
- Therefore, everything that happens in the universe is deterministic.
- If everything is determined, then we do not have freedom of choice, free will does not exist.

There doesn't seem to be any way around this conclusion. If things can only happen by logical processes, then everything must be deterministic. There is no other mechanism available.

Some philosophers would propose randomness / acausality to attempt to solve the problem, but the UP has already ruled that out, and it wouldn't solve the problem anyway. Making choices randomly, like flipping a coin, is really the same as not choosing.

Where does this leave us? It would seem we have no room for any free choice to happen. However, there is a "loophole" in the logic that seems to have escaped notice, and we'll get to it shortly.

First, it's necessary to define and frame the concept from the "outside", so we have some context. We need to know what free will looks like, we'll come back to how it gets made after that.

We need to consider:
- What choices are available?
- What mechanisms do we employ to make choices?
- What types of free will are there?

Two Options

To have choices there must be options. The most fundamental choice we have is between two predefined archetypal desires, and everything else follows from that. The two desires we can choose between are the Yin and Yang minds:

- **Yang, the desire to do "what's right".**
- **Yin, the desire to do what "feels good".**

Every living thing experiences / contains these two minds in some form.

The Yin mind desires material riches for itself. The Yang mind desires spiritual riches (e.g. "happiness") for all. These are the "two masters" we can serve, the two absolute "standards" or "rulers" we can measure things with.

Yin: Matter	Yang: Spirit
Free to not choose. To follow. Slavery. Yin desires to follow a leader	**Free to choose. To lead (oneself). Freedom.** Yang desires independence
Mind is the servant of the body	Body is the servant of the mind
Physical determinism Concerned with the past	Spiritual semi-determinism Concerned with the future
Apparent freedom to choose pleasures Many options	Only one right answer (But leads to "creative freedom")
Emotion, feelings, ego Pleasure and pain	Intellect, logic, reason Right and wrong

Urges of the body	Desires of the spirit
Seek pleasure, avoid pain	Seek right, avoid wrong
Short-term pleasure Long-term pain	Short-term pain Long-term pleasure
Wrong = choosing Yin (physical determinism) "Sins": *E.g.* sloth, gluttony *etc.*	Right = choosing Yang (spiritual determinism) "Virtues: "E.g. moderation.

This 1D level of choice is between "physical determinism" and "spiritual determinism", but the latter leads to another form of freedom, so we might call it "semi-determinism".

The imaginary man walking through the woods can choose which path to take, but he can't choose where those paths lead. He has a choice between two deterministic paths.

The left-hand path / Yin mind / slavery

The Yin mind is the mind of the body, the "reptilian brain". It is concerned with basic physical survival, especially in the context of being alone and "against the world", *i.e.* "in the jungle". It's also, somewhat paradoxically, the "herd mind", as the fear of being "alone against the world" drives their formation.

The Yin mind offers the freedom to not choose, and act on instinct or external instruction. This is where free will must begin because we start life as children who are not competent to make decisions.

Children must be free to follow until they are ready to lead (themselves).

Note that categorising the Yang mind as "leader" is not suggesting it is necessarily the leader of other people. The primary form of leadership is of oneself.

The Yin path gives us the freedom to follow the prompts of the external world reactively and passively, to be an actor on Nature's theatrical stage, reading out the script given to us by the "great architect" of the game.

It provides a set of predefined material desires which determine basic (animal) behaviour. It's the deterministic "simulated self", aka the "ego", which is generated by the host machine / "nature". The default position for all animals is to follow the emotions and urges of the body, without thinking about it at all.

This "mind of nature" offers the freedom to partake in all the wonders of animal life such as food, sex, family, triumph over the environment, and so on. Nature is the bittersweet experience of joy and suffering, pride and jealousy, victory and defeat. But it's like a scripted play in which you are playing a role written by someone else (e.g. algorithmically generated by the simulation).

The world of Yin is complex and paradoxical, it's beautiful and ugly, comforting and terrifying, vast and tiny. It's a "hell of a ride", like you're on a roller coaster, going where fate has determined, unable to change track.

The Yin mind is a "hedonistic" and selfish mind, seeing "pleasure" as the highest principle, focused on the material wealth and status of the self. It sees matter (physical wealth) as the highest principle, *i.e.* "God". The deification of earthly pleasures is a kind of slavery to matter, where your actions are determined not by your will, but by your body.

The right-hand path *Yang mind* freedom.

The Yang mind is the "mind of the parent", even the "mind of God", the ultimate parent.

Yang's desire is to do "what's right" by "absolute standards", and there's (often) only one right way to solve any problem, one right way to carry out a plan, one "absolute". There are many ways to do it almost right, or wrong, but only one "direct path". So, this path offers no real freedom either, at least in this perspective.

Yang is the archetype of freedom though, so it must lead to it. We've only classified free will as a duality so far, and there can't be much freedom in a one-dimensional object.

The emotional Yin mind

Every time I sit down to work on this book, my Yin mind (ego) raises various objections about it not being in the mood, or it being too difficult. It has all kinds of fears and counterproductive desires. It sees the work as a terrifying mountain to climb, and it would much rather browse social media.

I must force myself to work because my Yin mind is irredeemably lazy, it dreads toil and responsibility. Once I get started, all the negatives disappear and the work becomes worthwhile and enjoyable, but the Yin mind hates it and is always looking for excuses to do something else.

The Yin mind is like "entropy", it leads to sitting in front of the TV or social media. The Yang mind is the "organising principle", it leads to discovering virgin lands, conquering human frailties, and building great civilisations.

Of course, if I gave in to my ego and sat around watching TV, it would then start to berate me for being so lazy. It is a perverse and paradoxical entity, and it can be quite humorous.

The Yin mind is like a heroin addiction, The ego has complete control over the body's chemistry, and can induce various types of pleasure or pain to keep us hooked. It takes an effort of will to adopt the Yang mind, to give up the comforts of Mother Nature's teat.

Good / evil?

These are the two most fundamental options available to choose between, they set the underlying direction of all our other desires, and hence plans and so on. It's a very simple classification and does relate directly to the broad concept of good / bad.

The Yin and Yang minds are rather like the depiction we see of someone having an angel on one shoulder and a devil on the other, both offering their advice on what to do.

It may seem like the Yin mind has few redeeming qualities, but that would be an overly simplistic view.

In the context of morality, choosing the Yang mind is always "good" and choosing the Yin is "bad", but in other contexts it may be different. Both have their roles to play, and the Yin mind has skills and a perspective the Yang does not. In a survival situation, the Yin mind's paranoia, intuition, and rapid response is invaluable.

The emotions and feelings generated by the body are incredibly rich and complex qualia, they add all the colour into existence. Without emotion, life would be boring. Without the challenges we face, the "holes" we need to fill, life would have no meaning.

Ultimately, the Yin mind points to the Yang, like a signpost. All Yin properties point to their Yang counterpart. It is instructive in the extreme, if listened to fully. The "ego" is both our worst enemy and our best friend.

We must learn when to listen to which mind, appreciate their strengths and weaknesses, and find the appropriate balance.

Learning

If there is no free will, there is no moral responsibility because we don't choose our actions. If there is, the most important aspect of that freedom is being able to change ourselves. If free will is to mean anything, it must mean we have the freedom to choose to become "better people".

Whether we have the free will to choose between different breakfast options doesn't really get to the heart of the matter. What is important is whether we have the freedom to choose to modify our own desires and plans.

The result of any choice is a new "will". As we saw in the Turing machine, and in the EWAF process in general, it results in a new Fire / direction. But there is change happening in all elements, and this is the template for "learning".

The computation process involves writing new data (at Water), which corresponds to gaining new information. It ends in a new "will" which is also a kind of learning.

Decision making naturally includes the following changes, equivalent to "experience" / "learning".

Stage	Action	Result, "Learning"
Earth / Water	Gather facts, filter and sort	New facts are obtained
Air	Deduce or refine rules	Rule-set is modified + improved
Fire	Process rules and desires	Desire-set is modified + improved

Not Learning

We are free to choose to not-learn, to ignore data. Is that free will?

Our basic choice could be framed as "open / closed mind". The Yang mind is open to new information, Yin is closed. To learn, you must be open-minded, but Yin rejects information which makes it "feel bad". (Note, "bad feelings" are not prescriptive or causative. They push us in a particular direction, but we do not necessarily have to follow their command.)

The EWAF system begins with collecting and filtering data, but it's possible to entirely skip this step and base decisions only on existing data. We are free to reject all data, to "filter" all of it out, *e.g.* because we think we already know it.

In that case, learning does not occur. We can ignore data, but it leads to less information, thus less freedom of choice. So, it doesn't seem to be a route to free will, quite the opposite.

There is something interesting in the fact that we can choose to ignore data though. Computers don't do that. Instead of behaving rationally we can ignore important information. Instead of continuing a chain of logic to its natural next stage we can simply do nothing.

We can terminate a chain of logic with a "no".

We can ignore all the good reasons to do something and just say "no". When we ignore information we don't like, that is a choice between the Yin and Yang minds. We are choosing between either emotion or logic.

The act of ignoring things definitely feels like an overt choice. If I see something, then turn away and pretend I didn't, that is a deliberate act of will. I would have reasons for it, there is (emotional) logic behind the action, but a choice is certainly being made at that point.

To ignore data is to say "No, I will NOT learn!". This is Voice, the archetype of free will and of a logical NOT. That leads us to the next stage, the first mechanism our free will is expressed by.

The Origin of Free Will

If everything is determined by logic and causality, then there is only one possible route to having a free choice and that is negation. We can end a deterministic chain of causality with a "no" and create "uncertainty". The future then becomes undetermined. That is where the door of opportunity opens and free will is possible.

Free will is "holes in causality".

Causality is a process that occurs in discrete steps or stages. Left to its own devices it can only lead one way, to a single inevitable conclusion. The only way to "escape" this determinism is to end the chain of causality with a zero, a NOT, or a "no".

This is equivalent to creating a "hole" in causality, a break.

To escape determinism, you can only terminate it.

The only possible way a free choice can emerge is by ending the default chain of causality. In essence, we need a cause without an effect. (We can have a cause without an effect, but not an effect without a cause.)

A cause without an effect in this sense is perhaps equivalent to "absorption". If a ping pong ball hits a mountain, the cause has no (significant) effect. That doesn't mean the mountain exhibited free will, rather that it has high "mass" or "integrity". A human might ignore a provocation or temptation because he is the "bigger man" or has a "strong will".

Free will thus begins in the power to "not react", to "absorb" the impact. Reaction is deterministic. To "break the chains of causality", we can only negate them by not reacting. We might say you have to "turn the other cheek".

If someone attacks you (verbally) it can raise strong emotions and the initial reaction would normally be to attack them back; that is what the Yin mind tends to do. However, you have the choice to not react, to do nothing. If you allow a hole in causality to develop, other things will appear offering to fill it. You will find another way to handle the situation.

During normal development, we learn the ability to say no to any really bad ideas the Yin mind comes up with, and there is no shortage of them. We learn not to react blindly to temptations as we need a degree of self-control to function in society.

When we refuse to react to a deterministic prompt like an emotion, the possibility of acting differently opens up like a void. It gives us the opportunity to act on reason instead. We move from the future being determined, to it being undetermined.

Every time we say no to the Yin mind is like King Arthur withdrawing Excalibur from its dark, stoney prison just a little bit more.

Negation of determinism creates indeterminism

If we choose to give up a habit such as smoking, we negate the determinism of the habit by stopping it. We "just say no". What we do with all that spare time is undetermined, it's a hole waiting to be filled.

All life happens at the edges between holes and substance, Yin and Yang. Free will is an essential quality of life, and it is expressed as holes in a chain of causality. Events where a living agent said "No, I'm not going to react to that", or "I'm not going to continue this".

Stopping one causal process allows us to begin another. We cannot begin a new branch of causality as long as we're occupied with another, we must exit it and begin a new one. Saying "no" creates the hole, we then have freedom of choice in how we fill it.

The creative power of the void

When a chain of causality is broken a hole appears. Instead of time being filled with an endless process, there is no process. There is a lacuna, a lake to be filled with life.

When we have nothing to do, we become creative and find things to do, we start new chains of causality. At this stage we are free to create entirely new things, which have never been seen before. We obtain "creative freedom", and this truly is a form of "infinite freedom".

There is a creative power within emptiness. It's a "pull-force" like gravity. Holes desire to be filled, and "creativity" desires to fill them. "Boredom is the mother of creativity".

All the colour in life is created by the interactions between substance and "the void". Everything can be thought of as "holes in a continuum", where the continuum is "One".

Saying no to "Satan"

We might argue that boredom and "creative lacuna" can only be filled according to our pre-existing desires, and so they are deterministic, but we can also say no to our desires, and this is where morality links in. The "devil makes work for idle hands" they say, but we can always "say no to Satan".

The legal system is founded on the assumption that people have the agency to choose their actions. What this ultimately boils down to is, we expect people to be able to negate "evil" desires. We perceive that we have the agency to decide between "good and evil", and that happens in practice as "not giving in to evil".

The Yin mind is mandatory, we all experience its desires and sometimes they are bad. The traditional "sins" are pride, rage, jealousy, lust, gluttony, greed, and sloth. All humans, and probably all animals, experience these desires because they are built into the machinery of the body. We only become "good people" by learning to negate / ignore them.

We can say "no" to our Yin desires, and when we do that, we create the opportunity to get new ones. That is a creative process which offers infinite possibilities. We have gone from a single thread of determinism to an open choice with no predefined path.

Creative Freedom

The first level of choice is binary, one-dimensional. Creative freedom is another level of free will and offers choices in three-dimensions. You are free to create 3D things made of matter, and there are infinite possibilities.

There may be only two basic desire options, but desires only define the direction, not the path to take. Sometimes we must define the path ourselves and there are infinite unexplored possibilities.

Sometimes the path is undefined, and you have to make your own.

The imaginary walker we discussed earlier, at this level of conceptual reality, gains the freedom to "beat his own path" through the woods. He might decide that he doesn't want to go left or right, but straight on. He's free to ignore the predefined track and forge his own way through the trees.

When an engineer is designing a new product there will be aspects of the work where the path is undefined and there is no established "right" way to do it. In these cases, the creator must "invent a new algorithm" to fulfil the desire. We might call this "semi-determinism", in the sense that the outcome is determined but the path is not.

In a software project, the engineer (theoretically) receives a specification document ("spec") that tells him what the software must do, but it doesn't say how it should be done. The desire is defined, but the path is not. The engineer has a great deal of freedom in how the project is implemented, it's up to him to figure it out.

There is still only one "right answer" to the problem, the software must do what is listed on the "spec", but the route it takes to get there is up to its creator. Fire does not define the path, it provides the desire, the specification. The path (Air) is chosen by the individuals those choices concern.

The new path the walker beats through the woods is 3D, up/down, left/right, forwards/backwards. He has three avenues of freedom to explore, to "find his way" to the destination. Logic suggests certain routes, but he can say "no". The new path is undetermined until it is walked.

Greater mastery of a chosen profession leads to greater freedom in how it is expressed. A truly great artist or engineer can break all conventions to produce things of real novelty. Their highly developed free will allows them to reach the most "undetermined" end of the spectrum. No one could have anticipated their ground-breaking work, perhaps not even God himself.

"Nothing New Under the Sun"?

The concept of novelty is closely related to free will.

Do we truly have "creative freedom"? Have all things already been discovered? Is it possible for a human to discover or invent things that are truly novel, that have never been seen in the universe before?

While some modern inventions may be rediscoveries of things that have been lost and found many times over the course of history, it would seem rather arrogant to think humanity has thought of everything. The universe contains infinite possibilities, there's no way we could have been through them all.

Humans cannot create new fundamental archetypes, but we can create new mixtures. The Sex principle is the only mechanism available (at this stage of the UP) to extend the tree of logic. All the components of reality are defined "higher up", but we do get to mix them into new products.

If God is really seeking to "know himself", and we are the agents of that investigation, then we must be able to create or discover things that God did not know. God might "hope" or "suspect" things, but he can't have certain knowledge until the computation (thought) is complete.

When an artist begins a new painting, the desire is determined but the result is undetermined. It might be a picture of a house for example. There might have been trillions of pictures of houses created over the course of the universe, but the chances any two are identical is vanishingly small. The probability of novelty (in the broadest sense) is very high. (There must be a "scale of novelty" we could define...)

As suggested above, the greater skills someone develops over the course of their life, the greater freedom they have. A true master is able to create genuine novelty, which is presumably what the universe desires.

People as programs: "One seeks to know oneself"

If we consider individual people to be separate, parallel, thought processes going on in God's mind, then the people who do not "know God" (in a general, not religious sense), symbolise things that are still unknown / undefined about One. They are perspectives that aren't clear, are ambiguous, or conflict.

The UP seems to say that God "knows himself" by creation "knowing itself", *i.e.* by self-aware creatures understanding themselves. We are the "work" of One's thought process. We are (like) independent programs running on the universal computer, attempting to find answers to our/his questions.

People who are "in the dark" would symbolise questions in God's mind, and we're all in the dark to an extent, trying to understand our own being. None of us understand everything.

The only way any question can be resolved is by creative freedom, by discovering something new that was not already known. The only mechanism available to solve the problem is mixing / creation. We create new answers by finding a new mixture of thoughts.

A Map of Free Will

The fundamental archetypes provide a map of the principle of free will.

- One is the active principle, the actor, the "chooser", *i.e.* you.
- Yin and Yang are the two fundamental desires / minds / choices.

Yin and Yang are (like) the invisible angel and devil sitting on your shoulders, whispering advice into your ears. Here they also define the duality of determined / undetermined. Note that Yang may be "undetermined" in the sense of being the determiner, in which case the relationship is determined / determiner.

- The four elements categorise the stages we go through to make choices and are like nouns or "data stores".
- The three operators are the mechanisms which allow free choice and modify the nouns.

One - The Chooser, "Logos"				
Yin - Matter Yin mind: Determined			Yang + Spirit Yang mind: Undetermined, The Determiner	
Earth -- Results	**Water -+** Actions	**Air +-** Laws	**Fire ++** Desires	

	Sex Creative Free Will	**Heart** "Alternatives"	**Voice** Negation, "No"	

The process seems somewhat obvious in retrospect. It is perfectly intuitive.

Voice	NOT	Negation Terminate a chain of causality. The power to say "no".	This path is bad. I will stop following it. This is not my path.
Heart	OR	Alternation Seek alternatives.	Which other paths are available? I could go this way, or that way.

		The "power to discover others". (E.g. the "third person")	I could take this perspective, or that one.
Sex	AND	Combination Mix your own path. Begin a new chain of causality.	I will go this way, and then that way, and... I will mix this perspective with that one...

This gives us the overall picture, it's essentially a "map of free will". It tells us how it works. It identifies all the stages, mechanisms, and "data stores" involved. We can certainly add detail to it, but I think it covers all the basic principles.

What it does not do though, is prove we have free will. It tells is what it looks like and hence how we can describe and define it, but we still need to see if we can prove it's not just an illusion.

Picking a Card

Imagine picking a card for a magic trick. How do you choose one card out of 52?

The process will be as follows. We can use either the Yin or Yang minds to make the choice.

The Yin mind is the deterministic path.

There is only one possible deterministic path, and it is manifest as the instinctive Yin choice. It can give us what appear to be random choices, but they are always pseudo-random.

The Yin mind is reactive, fast, subconscious, and deterministic. It is the "default" mind. It will come up with an "intuitive" answer immediately which will be based on some subconscious logic. It provides the best answer it can, based on pre-programmed knowledge, rules, and desires.

Sometimes the Yin mind is right, and it's best to follow that first instinct. Not always though.

Choice 1: Negating the Yin mind.

We can either go with the reaction, or say "no", and at this point I suggest we break away from determinism. The immediate reactive response is the (one and only) 100% deterministic path, and when we negate it, we create indeterminism.

Negating the Yin mind creates indeterminism by terminating a chain of causality.

Note, negation doesn't guarantee that the answer we end up with isn't deterministic, it just creates uncertainty. We could decide to go back to our original choice, for example.

Also, we should bear in mind there is a spectrum of possibilities between determined / undetermined. Choices are complex things containing many factors, some may be more deterministic, some less so.

Anyway, if we choose to negate the instinct, then we "take control", we become active causal agents. We can use the Yang mind instead, although it is relatively slow and requires conscious effort.

Choice 2: Creative Freedom

Yang is the conscious mind. It can explore possibilities and look for alternatives. It can be creative and invent new ways to answer questions. You have the choice of which avenues to explore, and the choice of when to terminate the investigation.

This process is of undefined length. It can go on "forever" and contain unlimited sub-choices. It is "intellectual creative freedom" and it's fully under your control.

An artist may spend several years planning a particular work, and all the sub-choices involved will include a mixture of determined and undetermined factors.

Choice 3: Yang or Yin?

We now have the recommendations of which card to pick from both the Yin and Yang minds, and we can choose which one to enact. We start with a binary choice, then get a (relatively) infinite choice, then back to a binary one.

Indeterminable Choices

Determinists quite reasonably argue that our choices are driven by deterministic phenomena, *i.e.* prior knowledge, methods, and desires. We always choose things for a reason, and those reasons are developed by experience, logic, and genetics.

When discussing free will people tend to use examples which can potentially be determined by prior states, such as whether you could choose to have toast or cornflakes for breakfast. To attempt to reduce the complexity of the problem, we should consider choices that cannot possibly be rationally determined, such as picking a card from a deck.

Choices like this are indeterminable. There is no right answer. There are no facts or possible chain of logic which can determine the choice. The chooser can have no relevant desires to guide the decision. However, the UP suggests our first instinct of which card to pick is still deterministic, chosen by the subconscious / robotic Yin mind. Our route to indeterminism is to negate that instinct.

Negating an arbitrary choice

You can't prove your own agency to anyone else, but you can prove it to yourself, to an extent. You can observe yourself making the imaginary decision to pick a card from the magician's deck, then choosing whether to negate that choice or not.

Imagine the magician offers you the deck of cards, fanned out in his hand.

1. Choose one of the cards (but don't take it)
2. Decide, do you want to change your mind?

At this point, do not make a choice. Instead, observe what it is like to have that choice to make.

- Can you detect any reason or desire in your mind that is prompting a specific choice?

The answer will (probably) be no.

- Do you perceive yourself as having sole responsibility for the choice?
- Can you delay the choice indefinitely?
- Can you change your mind?

The answer to the above three questions will (probably) be yes.

Observing the void

That act of choosing to negate an arbitrary choice seems to be the "cleanest" form for examining our free will. We don't have any reason to go one way or the other. As we ponder it a void opens up. There are no prompts that can tell us what to do. Logic has nothing to say.

A choice like this cannot possibly be deterministic, there's nothing that could determine it, yet we can still choose.

It's worth spending a little time observing the character of "the void", it's a profound "quale" (singular of qualia).

As we wonder whether we will negate the card pick, it is impossible to imagine that we are not the agent. When we look for prompts, we encounter emptiness. We are alone. This experience of profound aloneness belongs to the One category, so presumably, God must feel like that.

The new chain of causality can only start with our decision, and we can take our time. Just as the universe could only start with One's choice. We suspend causality as long as we remain undecided.

As I watch myself going through this imaginary process it seems I have control over it. I am the "chooser", the active agent. I can observe myself directly, imagining the cards, choosing one, and so on. There doesn't seem to be any conceivable logical chain of causality that could determine whether I will negate the card pick. It seems I am the originator of the choice, an "author of causality".

(I think the only way this perception could be an illusion would be if this reality was an "action replay" kind of simulation, like a fully immersive movie, perhaps for education or entertainment.)

The Chooser

It's impossible to prove you are a conscious, causal agent to the outside world. It's only provable from the inside. We can personally know we have free will because we experience it directly as "the void". (Note, we have defined "knowing" as "personal experience".)

We might argue that even then it's impossible for me to know if I'm acting deterministically or not. My subconscious could be making the choice for me, and it could be predetermined.

However, in the example above, the negation of a card-pick, there is no conceivable deterministic mechanism which can give us an answer. It is a purely arbitrary choice. It cannot be predetermined by any logical process. We can observe the absence of prompts when we try to choose, we encounter the empty silence. No one will make the decision for us.

And then, when we get bored of staring into the void, we can observe ourselves choosing, essentially proving that we are the active agent, at least to ourselves.

Fear of the Void

The void can be frightening as it is the essence of "loneliness".

The sense of "aloneness" inherent in choice is significant. We sense that the choice is "up to us", there's no one who can decide for us, as much as we might like there to be. It can be a disconcerting feeling.

The Yin mind doesn't like to be lonely, and it doesn't want to decide things, so it avoids choice.

Choice requires work and responsibility, and the Yin mind sees no benefit in either of those things. Having to make decisions is difficult, and it means you must take responsibility for them, "where's the upside?" it wonders.

The Yin mind's "fear of the void" includes aversion to:
- being alone
- the unknown
- work and responsibility

These provide strong motivation to remain in the deterministic Yin mind and stay away from free choice. The void is where creativity exists, if we "avoid the void", we never break away from determinism and arrive at creativity. "Freedom lies beyond your fears", as they say.

The Yin mind desires a leader to make their decisions for them and to take responsibility when they go wrong. It is an "effect" but never a cause. It offers the freedom to be a follower, but not to be a leader.

To transcend the Yin mind, we must "stare into the void", and overcome our childish fears.

Parsimony

To attribute the act of choosing to something hidden like the subconscious is an unnecessary complication and conflicts with our perception. We can observe the active principle at work, and it is our own consciousness. We don't need any alternative explanation, and any candidates would necessarily be highly complex.

The UP correlates "the chooser" with consciousness under the One category, and choice under Voice. It says choice is an act (mechanism) of consciousness and only consciousness can make choices, and that is how we perceive things to be.

The simplest interpretation of the facts would be that our perceptions are accurate, and we are indeed the authors of our own fate, albeit within the constraints imposed by nature.

We cannot avoid responsibility for our own choices. God didn't make us do it.

Data is not instructions.

The determinist argument sometimes falls into a category error, in which the property of agency (One) is attributed to "rules" (Air). Logic or "reasons" can't conceptually "choose" anything.

Information can lead to choices, but it can't make choices. You can have all the reason and desire in the world to make some choice, and still not do it. The act of deciding is a different class of thing from facts and desires.

A list of logical arguments cannot make a decision. It can indicate which is the right choice, but it's a passive concept, a list can't do things. Conceptually there must be a "chooser", an active principle that corresponds to the One category.

In a computer, decisions are made according to the program, but it's not the code that actually makes stuff happen, it's the processor. The conscious decision maker is the only possible arrangement which can account for our observations of reality (i.e. ourselves). There is no other explanation, and that is a logical proof of its existence.

True randomness?

Working through this topic has brought up the possibility that the claim that randomness does not exist might be wrong. After thinking all of this through it seems that true randomness (although not acausality) may in fact be possible, but its source must be an intentional act of consciousness.

It seems it is possible for us to pick a genuinely "random" card using the Yang mind. We are able to make choices with "no mechanism". It is not deterministic, there is no logical cause. The chooser is able to choose even in the absence of prompts. This does seem to be at least approaching true randomness.

It implies the Yang mind can (maybe) create truly random data, although it needs a bit more thought. If that's true, then how does it do it? It is a complex system, perhaps it's always pseudo-random, I'm not sure.

So, I might need to add a caveat to the phrase, such as:

True randomness does not exist, except as the intentional product of a mind.

Energy and Determinism

I have already described the Yin and Yang minds as akin to the duality of entropy / the organising principle, and we can usefully link them to concepts of energy in other ways.

Given that the Yin mind's answer is instant, and takes very little energy, while the Yang is the opposite, we might propose that the more energy we put into the process of choosing, the less deterministic it is. As mentioned above, the duality determined / undetermined is a continuum. There can be degrees of determinism which can be quantified.

It takes energy to break out of determinism, to make truly free choices, and the more energy you can put into a choice, the "freer" / less deterministic it can be. A more thorough investigation takes more time but provides more information, which leads to more freedom / less determinism.

That seems an obvious analogue of "more money leads to less slavery". The more energy / money / information we have, the freer we are to do things.

The more energy a system has, the less "predictable" it is.

The Yang mind leads to creativity and invention, it acts to increase order and save energy in the long-term, for many people (e.g. family). It leads to material "safety" / security / wealth. The Yin mind essentially only consumes energy, instinct has no creative power. It acts to save personal energy in the short-term leading to stasis or regression, and insecurity.

The Yang mind is like the organising principle, creating order, making novel energy saving inventions. The Yin mind just consumes energy, like entropy.

The Gravity of Yin

gravity
- weight, importance, significance, seriousness
- from Latin gravis (heavy), note "gravid" (pregnant)

It takes effort to think, and nature loves to save energy. When we try to solve problems and make choices there will be a natural tendency to be "pulled back" toward determinism, to save energy in the short-term and take shortcuts.

Yin hates change. When we try to change our habits (perhaps "programmable subconscious"), such as giving up smoking, it can be very difficult, and we often end up back where we started. A habit is like a hole. Stopping a habit is like climbing out of a hole, you might stumble and fall back down to the bottom again. That is where gravity wants you to be.

Yin wants Yang to be at the bottom of the hole, in determinism, and that desire is the "pull force", the archetype of gravity and other "attractive" forces. Matter always seeks the lowest possible "energy state". It wants all holes filled, all jobs done, all facts explained.

It manifests as a pull-force, but it's illusory. It's always caused by a push. In the case of choosing, I think the push force is the energy it takes to use the Yang mind. It takes effort to forge your own way through the woods, and we get tired of thinking and go back to following the beaten path.

So, it's the effort required to maintain a push force that we are producing that causes "gravity". I assume this analogy is significant. We must work to move away from Yin, and if we don't have enough energy to escape the "gravity well", we fall back down to the bottom again.

Gravity is an expression of the "maternal" nature of matter, desiring to pull its children closer, to be more like Yin. It's like a "love of the past", resisting change and fearing loneliness. "Inertial mass" resists change. It takes work to move matter.

Enlightenment (learning / freedom) is like walking up a mountain, away from the determined / known Earth, towards the undetermined Sky. All the while, gravity seeks to pull the climber back down; he must constantly work to increase his elevation and be wary not to stumble.

Alchemy

When I began this analysis, I didn't have any preconceptions. I wasn't looking for Alchemy, but I found it nonetheless. At least I assume that this is what Alchemy was based on, it seems to make sense. The investigation which led to this theory, as documented here, potentially stands as a rediscovery of the foundational concepts of Alchemy from the first principles of science. That's pretty amazing.

This strongly suggests that the UP, as described in this book, is an accurate description of the conceptual framework, because it seems that it has been discovered before, perhaps many times.

Given its solid foundation in observable fact, it seems that Alchemy is / was always a viable and legitimate approach to science, and arguably the right one.

I don't know the original source for this image, but it shows a set of alchemical relationships, and the structure corresponds to the structure of the UP. The names don't really matter, the relationships do. It depicts the ten entities, and their relationships.

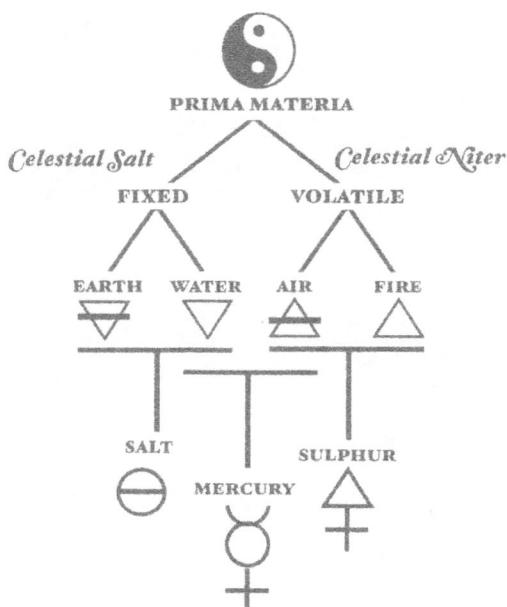

The structure below was discovered independently, and without prior knowledge of the above. I didn't go looking for it, it just emerged from observation, as explained. It is the same structure.

Level 1: One, unity The Executive. Origin. Source and Destination. "Alpha and Omega" "I am". Consciousness.				
Yin - Matter, solid objects Absence, shadow. "I am not"			Yang + Spirit, abstract objects Presence, light "I am"	
Earth -- Results. The product. Value.	Water -+ Actions. Cycles		Air +- Laws. True/false	Fire ++ Desires
	Sex Combination. Mixing	Heart Alternation	Voice Negation, Reflection, Logic.	

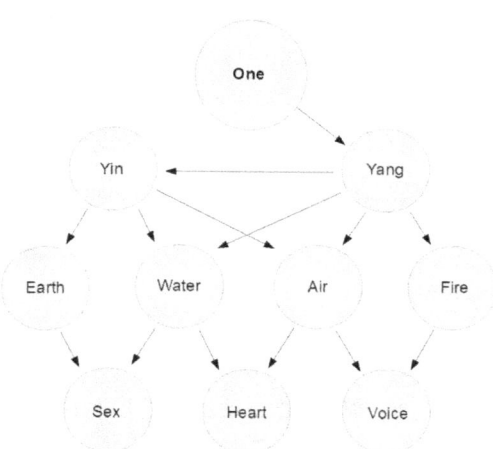

What Is Alchemy?

The common view of Alchemy is as a misguided material science. A search for "alchemy" on Brittanica.com returns:

- alchemy (pseudoscience)

```
https://www.britannica.com/search?query=alchemy
```

This does fairly represent the modern view, but it seems a bit uncharitable to label it as such. Why not "proto science" or something? It's almost as if

Brittanica wants to be sure people dismiss its ideas before they've even been examined. Most modern sources seem quite certain alchemy was a primitive delusion with no real merit, but this investigation has provided reason to think otherwise.

Alchemists are said to be concerned with obtaining the "philosopher's stone" which would allow them to "transmute lead into gold" or obtain "eternal youth".

alchemy
- a medieval chemical philosophy having as its asserted aims the transmutation of base metals into gold, the discovery of the panacea, and the preparation of the elixir of longevity

https://www.yourdictionary.com/alchemy

If the UP is correct then, far from being pseudoscience, alchemy was the historical foundation of science for good reason. It is the only conceivable theory of everything, and it unifies all perspectives.

Importantly, the concept of "enlightenment" in science and religion is unified under the UP. Both are attained via the same path: observation and reason, "logos". Perhaps the opposite of a "mystery school", the UP eliminates the "mystical" by explaining it. It provides a logical structure into which all concepts fit. It brings the "spiritual" into the realm of science.

The conclusion must be:

Alchemy was an "enlightenment school".

Science of the Golden Age

The basic parts of alchemy (duality, the four elements, the seven principles) are found all around the world. Most, if not all the major religions contain some reference to its parts. They appear in the Eastern, Abrahamic, and "Pagan" religions in various forms. Alchemy is perhaps a bit like "Osiris", the Egyptian god who was cut into many parts and scattered around the earth.

So, how did they get there? How did the parts of alchemy end up spread across the world, embedded in all these different religious systems? This is necessarily rather speculative, as what happened before our recorded history can't truly be known, however there is a common "myth" of a "golden age" of humanity in the past.

How reliable is our recorded history? How long ago was this "golden age" if it existed? It's hard to say, and this isn't a history book. However, if there was indeed a time in the (not too distant) past when a great civilisation arose which spread around the whole world, then that might explain some oddities in ancient history, mythology, art, and architecture.

I speculate there was indeed a "golden age", that alchemy was the combined religion and science of this world-spanning civilisation, and that it was in fact the reason for its success.

Alchemy was the unified science / religion of the "golden age".

The principles of the UP are a solid foundation, not only for science, but for all human endeavour. Alchemy / the UP is tantamount to "natural law", which is Yang / "God's law". I think a civilisation that "worshipped Truth and Reason as gods", would be a great civilisation. At least, I'd like to live there.

In this view, it would have been the abandonment of natural law that led to the end of the golden age. Which is equivalent to mankind "rejecting God's law" and making their own. The "ages" of man can be defined by their laws.

Hidden Origins

I suggest that the depiction of Alchemy as a materialist, money-oriented, endeavour was concocted to hide its true intent. History tells us that when Christianity took hold of the west, the "pagan" religions (i.e. all other views) suffered considerable persecution.

The UP could certainly be the basis of a religion. It deals with concepts of "God" and "gods", and it contains an innate moral code, "natural law", which is the origin of what ended up as "common law", and also the "jus gentium".

`https://en.wikipedia.org/wiki/Jus_gentium`

I suggest that Alchemy wasn't just the predominant science before Christianity but was also its religion. It may have lost its place, not because of a lack of virtue, but the opposite. It was a superior way of viewing the world than the Abrahamic one, and as such was a threat to the hegemony they sought.

Alchemy was disguised as chemistry by its sages to try to preserve their knowledge in the face of what could generously be described as "extreme censorship".

What heresies would they have been committing? Probably all of them. Assuming I've interpreted it right, the UP leads to many "heretical" conclusions.

Other Gods

One is the highest God (capital G) who exists outside nature, but other gods (archetypes) exist within nature, and God has delegated power to them. "The gods of nature", like all creatures, have free will.

The UP can view nature (duality) as a deity with a mind of its own, it is the "Great Yin", the "Goddess". So, God has a "wife", and it's the universe *"creation"* the "dream". The UP describes the existence of "gods within nature", such as the four elements. They are individual entities, minds with their own perspectives and powers.

It says God allows things to run themselves and he doesn't need to interfere. He essentially allows "lesser gods" to determine mankind's fate (because he can trust them). The God of the UP doesn't seem to be "jealous".

I think the way free will works in this case is the "gods of nature" can't (practically) choose to go against God's will because that would be to go against their fundamental nature. They are free to choose how they enact his will though. They have "creative freedom". God provides the "why", they determine "how".

Nothing is free to change its fundamental nature, not even God. Which is probably another heresy.

The historical record of people's beliefs is very complex, but the nine entities below One are the original "gods of nature" and were probably the prototypes of "pagan" deities and pantheons.

Non-Omniscience

The UP suggests that God doesn't know everything. It seems he doesn't even know himself as the universe was apparently created to answer that question. Although maybe that question is now answered and there are others to consider? However, if he didn't already know the answers, then God does not know the future.

If the universe is a process of God finding things out, then he is not "outside time" or "omniscient".

Young children tend to view their parents as "gods" because they are so Yang to them. We imagine our parents are like "super-heroes", but as we grow up, we discover they are more limited than we thought. Maybe those conceptions of One are like this, an "immature" view of his true nature.

The UP says "God" is "good" and "perfect", "omniscient" and "omnipotent, but only within the realms of what is logically possible. God cannot "defy logic", he has limits and he does not know the future.

The UP says no one knows the future, not even God, because to "know" a thing you have to have experienced it. We can only "know" things that happened in the past, by definition. Yin is "known", Yang is "unknown".

Non-Omnipotence

God cannot be "omnipotent" in the way that it is commonly defined. One, for example, could never "create an equal to himself". There are things God cannot do, he is "bound by logic".

It seems to be inconceivable to have any other system of logic than the one we do. That might be just how our minds are made, but if duality was the only possible route available to create the universe, that would constitute a constraint on God.

Apotheosis

The worst heresy of all however, would probably have been the ultimate goal of the religion, a belief in "apotheosis", the transmutation of "man into god", aka "lead into gold".

apotheosis
- the transformation of a mortal human into a state of divinity and immortality

The belief was that a man can become a "god" (i.e. exit the simulation, "win the game") by their own efforts. This would be anathema to Christianity. If you can (only) save yourself, why would you need Jesus?

Reading between the lines, Alchemy's intent must have always been personal spiritual enlightenment, to the point of apotheosis (note: that's quite a common claim). Its aim is to transcend the material world. To leave the "wheel of incarnation" and to go to "Heaven".

The mechanism of enlightenment, suggested by the UP, is personal virtue obtained via "reflection" (the "straight path" *Voice* Christ principle).

The obvious problem with that is the Christian church also claimed to know the route to Heaven, and the two do not concur. It claimed belief was the key, not virtue. Alchemy's method was based on reason and personal knowledge ("gnosis"), whereas the Christian one is based on faith.

The two religions would have been competition for minds, and Alchemy had a superior product.

**Alchemy was teaching apotheosis via personal development.
It was an "enlightenment school".**

Alchemy taught that anyone could attain a state of immortality by their own efforts, at least this seems to be the true message encoded in the symbolism. It taught that the path to heaven was by personal virtue, rather than by "faith". The natural conflict with Christianity is clear.

Physical transmutation

Not all Alchemy is concerned with the spiritual, it also deals with matter. I would tend to believe that it is possible to transmute physical elements, and alchemists in the past may well have done so. Given the UP as a basis for science, alchemists would have had a significant advantage over modern day scientists, a "top-down" view of physics which is lacking today. "Lead and gold" sound like allegories for Yin and Yang, or "man and god", but could also be literal.

Maybe its secrets do also reveal how to transmute physical metals. The material world reflects the spiritual one. If a man can be turned from lead into gold metaphorically, then maybe lead can be turned into gold literally, by (a material analogue of) the same process.

This presumably where the "dissolve and coagulate" cycle originates. To obtain enlightenment we must dissolve the old, bad ideas of the Yin mind and the "coagulate" a more Yang version. This cycle is repeated until we wake up.

The Philosophers' Stone

What is the "philosophers' stone"?

Wikipedia says:

The philosophers' stone ... is a mythic alchemical substance capable of turning base metals such as mercury into gold... It is also called the elixir of life, useful for rejuvenation and for achieving immortality; for many centuries, it was the most sought-after goal in alchemy.

The philosopher's stone was the central symbol of the mystical terminology of alchemy, symbolizing perfection at its finest, enlightenment, and heavenly bliss. Efforts to discover the philosopher's stone were known as the Magnum Opus ("Great Work").

https://en.wikipedia.org/wiki/Philosopher's_stone

The philosopher's stone is generally portrayed as being made of matter, for the purposes of transforming matter. But philosophers don't work with matter. Builders and mechanics work with matter, philosophers work with ideas.

The motivation for the "Great Work" is supposed to be to make gold, and so become very wealthy, but does a philosopher really desire worldly riches?

philosophy
- "love of wisdom"

A love of wisdom is not the same as a love of money. In fact, they're somewhat opposite. We might suspect they are a duality. You can't "serve two masters", you can't seek truth and money with the same will / spirit.

What philosophers want is knowledge, truth, wisdom. They want to be able to understand things. That is the prize the philosopher seeks. The greatest fear of the truth seeker is to live and die, never truly understanding anything. To spend a whole life trying to fathom reality only to fail is a terrible fate.

Information is like energy, as discussed. It allows you to do more things faster, and time is really the most precious commodity of all.

Money can't buy the things that really matter to a philosopher. It can buy material freedom, but truth can buy understanding and contentment, a type of freedom with far greater value. It may even offer freedom from the simulation itself, the ultimate prize, true freedom.

The most valuable thing in the world is absolute truth. It's far more valuable than gold. If the philosopher is seeking truth, then that is what the stone must be made of.

The philosopher's stone must be made of truth

The stone must allow understanding, it must be a foundation-stone for knowledge. The one thing that people lack in this world is a solid foundation of truth on which to build. The world is full of theories, beliefs, faiths, and opinions, but they are all built on the shifting sands of assumptions.

What the philosopher, the scientist, and the religious person really need more than anything is a solid rock of hard facts to build their understanding on. Only this way can they be sure they are on the right track. This is what the UP provides, a firm foundation on which science can be built.

The UP can unify all perspectives. There can be a science of spirit which is just as rigorous as any other, if not more so. The UP is the unification of science and religion and of all disciplines.

For these reasons, and for all the other reasons described in this book, I think:

The Universal Plan is the real philosopher's stone.

1. It's obtained by doing philosophy.

The UP exists "out there" in reality, freely available for anyone to "see" / deduce. None of it is hidden, but you do have to actively look for it.

The only way to understand the UP is by thinking it through personally, by doing philosophy yourself. There's no short-cut, no way to obtain it other than by following the reasoning. It's not enough to read the words, thinking must be done.

Understanding the UP is, just like the universe itself, an iterative process. It requires practical philosophy, *i.e.* thinking for yourself. It takes multiple cycles of "dissolve and coagulate" to grasp the information it contains.

2. The UP structure is, conceptually, the most solid / hard thing that exists.

The UP is the universal template that never changes. It's timeless and indestructible. None of its parts or their connections can ever be broken or changed, it can only be built upon.

It is the "Great Yin", the most solid "matter" that exists. It is the foundation everything else is built on and from. It is like an infinitely hard stone which can never be changed or turned into sand by the actions of time, unlike a stone made of physical matter.

3. It's a foundation-stone.

The UP is the only solid, complete, logical body of information available to build a system of science on. There is no other. It's the "only game in town". It's a rock on which you can build a "castle of knowledge". Without it, anything you build could come crashing down at any time.

It's maybe also the figurative "stone the builder refused". Modern science rejected Alchemy and the idea of the four elements some time ago.

Matthew 21:42
Jesus saith unto them, Did ye never read in the scriptures, The stone which the builders rejected, the same is become the head of the corner

4. It's the foundation of a complete theory of everything.

The UP contains all knowledge in the universe, in seed form. Everything that exists is made according to this template. Having it is a huge head-start for any analysis or investigation.

Trying to understand the universe from the perspective of physics is like trying to put together a jigsaw when you don't know what the finished picture should look like, but the UP is that finished picture. It's the blueprint of the universe. This is what the jigsaw looks like when it's all assembled. All we now must do is figure out how all the pieces we have fit into it.

The UP contains all the information on how nature works, somewhere. Finding out how to apply the rules to specific cases in material reality will remain hidden within the vagueness inherent in the system until it's teased out. "The devil is in the detail", as they say.

Infinitely Hard, Infinitely Meaningful

The component parts of the UP are bound together with a potentially infinite number of interconnections, just as musical notes can be combined into an infinite number of melodies. As the world of concepts grows out of it, and more connections are made, the whole structure becomes more solid.

The atoms in a physical material can only be connected to their immediate neighbours, so its strength is finite. No matter can be infinitely strong. Here, every concept is connected to every other. These connections are like the cement which binds the components together, but they aren't constrained to connecting only to their immediate neighbours, because they're not physical objects.

This could be viewed as making the UP infinitely "hard", conceptually speaking, *i.e.* not susceptible to being changed by anything in the material realm or by time itself. This would be equivalent to being "infinitely true", as it's a set of concepts and principles.

All those connections also make meaning. An infinite number of connections would be infinite meaning, so the structure is also "infinitely meaningful".

What a philosopher wants and needs more than anything is a solid foundation for their understanding of reality. A philosopher wants a rock of solid indisputable truth to build their analyses on. He wants to understand the foundational truths of reality. That's worth far more than gold to a seeker of wisdom.

This is what the UP is, it is that foundation. It's the "master-key" to understanding the universe.

The Sun

The UP is perhaps the archetype of the physical sun. The UP could be seen as the "sun behind the sun", or the "invisible sun", as it's visible only to the mind's eye. It provides the "light" of information and truth, and just as it's much easier to work in the light of the sun, it's easier to work in the light of good information.

The Sun	The UP
The provider of light and life	The provider of truth (light) and life
Exists in the physical sky	Exists in the conceptual sky
Centre and foundation of the solar system	Centre and foundation of the spirit-world.
Inspires awe by its brilliance. Allows people to see things	

The "Alkahest"

A suggestion regarding the identity of the alchemist's "alkahest".

In Renaissance alchemy, alkahest was the theorized "universal solvent". It was supposed to be capable of dissolving any other substance, including gold, without altering or destroying its fundamental components.

https://en.wikipedia.org/wiki/Alkahest

The alkahest must be logic. It is the "universal solvent" which can break anything down into its components without changing it.

Final Thoughts

In this chapter I want to offer a few miscellaneous thoughts and observations about the system and tidy a few loose ends.

The Perfect Machine

The UP is like a template or master plan which describes the fundamental components of reality and relates them together as a "universal language". It is a single tool that can make any conceivable thing; it's the "perfect machine".

The Genius of the UP

This system appears to be a work of extreme genius. To devise a system this simple, which can build universes with just ten archetypes, would take quite a mind to say the least.

Note that it is also "self-documenting". It explains itself, why it exists and how it operates. It is both the machine and the user-manual, all rolled into one. It is unfathomably concise and efficient.

It would seem there is no way the UP could have come into existence other than by deliberate intent and considerable intelligence; I don't think humans could invent anything this clever.

I believe this is good evidence for its validity. I certainly am not bright enough to have devised this system, I merely stumbled upon it, like a monkey surprised by his own shadow. It has shown logical consistency throughout all the investigations so far and has already achieved quite a lot in explaining some important concepts.

However, we do still have the question of whether it was strictly "designed" to explore. There is an intriguing mystery there, or at least something that needs a definition.

The Limits of the UP

There are some significant limits to what the UP can tell us.

The UP is only a beginning

You would be forgiven for thinking that a TOE should more or less immediately solve all problems, explain everything, and provide universal abundance and world-peace, but sadly this is not the case. The framework is fully described by the UP, but this is only the beginning, not the end of science.

Rediscovering the UP was relatively easy because the most Yang things are the simplest. Physical matter is very Yin, and so it will be more complex. There's plenty of work ahead.

The UP doesn't provide all the answers, but it does provide a well-defined system of categorisation into which all the answers fit; it acts as a rigid guide to help find them. It provides a solid foundation on which to build knowledge, but we still have to do the building.

The UP can only define what is conceivably possible.

The UP only describes concepts and what is conceivable. It assumes that everything is made of logically constructed ideas, but that assumption could be wrong. perhaps reality can do things which are not conceivable. If that was the case though, all science would be equally wrong, so the UP would be in good company.

The UP can only define things from our frame of reference.

The UP reaches its limit with the concept of "One". It cannot describe what exists outside the universe. It can hint at what should be possible, but the universe is self-contained and we cannot see out of the container. There are things about One we cannot know, and this is the uncertainty which allows for it to be nested, hence the possibility that this reality could be a simulation.

Relativity

An important limitation of the conceptual framework is that things can only be described in relative terms.

At least some of the Yang properties must be considered as relative as opposed to absolute. Yang tells us what One is like, but Yang can only be described in relation to Yin. All Yang's attributes can be rephrased as "- relative to Yin". For example, Yang says God is "eternal", but we perhaps should strictly interpret this as "eternal relative to the universe".

Yang says, "God is mind, and has no body", but again, this must be interpreted strictly as "God has no body relative to the universe". It really means that if God did have a body, it would be "irrelevant" to creation. We simply have no access to that information.

Vagueness

Another limitation of the theory is its broadness means it can be quite vague. We must "interpret" and "abstract", to think in extremely general terms. It is primarily a qualitative theory and will always be so.

However, I'm sure it would be possible to formalise it and express it in some logical system, and eventually to code it into a program that could be run on a computer.

At some point, the UP needs to be applied to logic itself, to determine how exactly it is structured, and how it should be expressed. That would then need to be correlated with mathematics. It would be quite a big job but is ultimately necessary. That should help to remove most of the apparent vagueness.

A new start

If this theory is correct, it would be fair to say that science hasn't even really started yet. The science we have today is largely founded on misclassified concepts. It assigns the wrong properties to the most elementary phenomena. (e.g. "spacetime", "infinity", "random").

The UP is the only conceivable solid foundation for science. Without it, it's all just "castles built on sand".

Physics assumes there is no connection between "matter" and "meaning". The materialist view is that the universe has no ultimate meaning or purpose. This means that physics can never actually explain anything. It cannot join matter to the world of concepts (which it assumes do not exist).

Modern physics is only capable of telling us HOW things work, it can never tell us WHY.

The UP, on the other hand, connects matter directly to "meaning". Matter is a symbolic representation of meaning; matter is what matters. It explains exactly what the relationship is. It also shows us that information is a form of matter, joining the concepts at a deep level.

The UP can provide rich and detailed explanations which describe WHY things work the way they do. Modern materialist science cannot do this.

Only Three Dimensions?

The UP describes explicitly how the three-dimensions of space are formed, although it does it in such abstract terms, the physics is not immediately obvious and an analysis will have to wait for another day.

It clearly makes the prediction that the maximum number of dimensions required to describe any fundamental system is three. While systems can be stacked in layers to create the impression of more dimensions, if the layers are looked at individually three is the maximum. This is another bold prediction, and only time will tell if it's correct.

Nature always finds the most elegant solution to problems. Perhaps the universe can achieve everything it needs with three and there's no need for extra dimensions. Perhaps the most important impact of this conjecture, should it be true, is that it means nothing in the universe is beyond human comprehension or imagination. We can't imagine a four or five-dimensional space, so if space had those extra dimensions, it would forever be beyond our understanding. Theories that space has more than three dimensions are necessarily a bit "mystical".

However, if everything works in 3D, we should have no problems visualising and understanding it. It makes full knowledge of reality possible for humans.

Having said all that though, the UP structure contains ten entities which might appear as dimensions in some systems. I'm not sure. If we consider our senses to be dimensions, then we have at least 14 of them. This is an open question, it needs testing.

The Fifth Element

It's believed in some traditions that there is a fifth element, sometimes called "aether" or "quintessence". Aether theories in physics were and still are common.

quintessence
- the pure, highly concentrated essence of a thing
- the fifth and highest element, coming before air, earth, fire, and water

From a physical perspective, it's the medium light travels through.

(luminiferous) aether
- the physical medium of light and electromagnetic phenomena

How does this fit in with the current theory? Well, there is absolutely no room for another element in the list of four. There cannot be a fifth element which is a counterpart to the other four, *i.e.* on the same level of the hierarchy.

We covered the answer to this question in the section on physics, but just to recap.

The "fifth element" is simply the parent-category, one level above in the nested-list. It's the container. From our perspective, the aether is "space".

The "aether" is what everything on a given level of reality is made from.
- One is the "aether" of duality because it is the parent category.
- Duality is the "aether" of the four elements, and so on.

"Aether" is a relative term in this view.

Plasma is the "aether" of the phenomena it hosts like lightening, corona effects, the "northern lights" *etc.* (Physical) water is the "aether" of water flow patterns, turbulence and so on, again a complex family of physical phenomena.

The aether is the host of the phenomena it supports, and it's a one-to-many relationship.

Our aether is "space"

Modern physics contends that there is no "luminiferous aether", and that light travels as photon particles. The UP says the aether is a necessary concept and light is made of waves. All information is made of holes in a continuum, equivalent to a "wavy line".

The aether is the same thing as "space", it's the venue, the host, the container. The Fire element is zero-dimensional, so anything above it in the hierarchy must also appear to be 0D which is effectively "hidden". The aether / space appears to be "nothing".

We see properties in material things located in space, and space seems to have none, but this suggests "space" is the origin of all those possible properties. It has them, but only in potential form (Yang). The things made of matter are then realisations of the potential provided by the container / host.

We see things as having properties, and space being an absence of them, but the properties we attribute to matter are actually features of the aether that matter is using / symbolising. Aether is like the hardware; matter is the software.

Fractals and Nesting

One of the most important tasks ahead in refining UP theory is to establish what it's natural rules of nesting are. I can only offer some suggestions at this stage.

We could depict the UP as a list, as follows.

```
One
- Duality
- - Fire
- - - Voice
- - Air
- - - Heart
- - Water
- - - Sex
- - Earth
```

The question is, how does this expand out to allow for more detail? How do we build stuff with it?

4E Nesting

To provide a finer, more detailed description for concepts, we could create a nested list of the four elements. For example, the Water element contains all the following archetypes / concepts: emotions, force, action, strength, power, waves.

To separate them all out we could divide them into a nested set of four as follows. Here Water is the "quintessence *aether* fifth element" of all the sub-elements. *I.e.* "Water.Fire" is a sub-category of Water.

Water.Fire - Emotion (motivation) - "why"
Water.Air - Strength, power, force - "how"
Water.Water - Work, action - "who *where* when"
Water.Earth - Waves, circles, cycles. - "what" / form.

This is probably a useful way of doing this in some cases, but I think it would be an "approximation". I think the UP probably nests as a whole, like a true fractal.

Whole nesting

The proposition here is that every category in the UP contains a copy of the whole UP.

For example, if we focus in on the Fire element, what does it look like inside?

It will look like a copy of the whole UP, where "One" is "Fire". *I.e.* Fire takes the place of One in the list.

```
Fire (One)
- Duality
- - Fire
- - - Voice
```

- - Air
- - - Heart
- - Water
- - - Sex
- - Earth

Here there is a Fire.Duality category, a Fire.Fire category, and so on. The Fire element has its own "Yin and Yang", its own "Heart" and so on.

Depending on the dimensionality of the system / concept being defined, we might need to assume we'll need to account for up to nine subcategories during the analysis.

Nesting rules and DNA

The nesting rules we discover would presumably be an analogue of physical DNA. The method used to create the family-tree of the conceptual framework should be analogous to the method used to create biological life, although they will probably be reflections of each other, and so opposites in some ways.

Space / Time Units

I'm leaving any detailed discussion on physics to another book. Matter is complicated and needs careful thought. There is however one thing that comes out of the principle of duality that is simple enough to include, some SI units may be redundant.

https://en.wikipedia.org/wiki/SI_base_unit

There used to be a website at blazelabs.com that had some interesting ideas on physics, including a re-working of all the SI units in terms of just two, space and time. The fact that it's possible to do this suggests it's what we should do, at least some of the time. It's still available on the Wayback machine.

https://web.archive.org/web/20211016044745/http://blazelabs.com/f-u-suconv.asp

I believe this system was originally proposed by Dewey B Larson.

http://www.lrcphysics.com/

It seems all the basic physical relationships currently expressed in the SI units like mass, inertia, inductance *etc.* can be written solely in terms of Space and Time. If it is correct, then this is the system we should use, at least for theoretical work, because it's more fundamental and hence closer to the truth. If the other units aren't fundamental then they will inevitably hide information from us, like a filter.

Space and Time correspond directly to Yang and Yin. Space/Yang is distance *difference, it's "active" because distances can change, and it's a real thing we perceive directly. Time*Yin is a "just" a measurement of changes in space, it's dependent on the existence of space.

Really all there is then, is space which is changing. Space is "distance", the same thing as "relationship", so reality is made of "changing relationships".

3D Time

We have three-dimensions of space, and in this view, there are also three-dimensions of time, existing in the present. Space / time is a complementary duality, and time is a "reflection" of space.

They are most simply expressed as velocity, acceleration, and jerk, the three derivatives of space. This may seem like a strange way of looking at it, but it makes good sense from this perspective. If we view time like this then it becomes a perfect mirror image of space (conceptually).

The three dimensions of space and time are as follows. (^ = to the power of)

Dimensions	Space (metres)	Time (seconds)	Space / Time
1	Length (m)	Change (s)	Velocity - m/s
2	Area (m^2)	Change of Change (s^2)	Acceleration - m/s^2
3	Volume (m^3)	... of Change (s^3)	Jerk - m/s^3

Jerk

This theory argues that acceleration is a different "dimension" from velocity in the sense that it's a different category of thing and therefore has different properties and effects. For example, a charged particle moving with constant velocity does not emit light but one under acceleration does. What the effect of "jerk" is, I'm not sure, but there must be one which is distinct and measurable.

I think there are two possibilities.

- The present physics might be wrong, acceleration may not be the deciding factor, and light may only be emitted under sufficient jerk
- If present physics is correct, and light is emitted under acceleration, then some other effect should be noticeable under jerk.

The UP implies that the purpose of light is to convey information to the universe about a change of state, and a "jerk" would presumably require two items of data to identify it. I don't know how that might manifest.

A Map of Physics?

If we assume there are only 3 dimensions of time or space, we can build a table to show all the possible combinations of relationships as follows. This table then defines every possible fundamental relationship that can exist in physics. It's (potentially) like a masterplan or overview of physics.

Some of these relationships don't seem to be known phenomena, but I must hold my hand up and admit that I'm just a programmer, not a physicist or a mathematician. I may get round to working on this in due course, but it's not my forte and help would be welcome.

Some known relationships like density (T^3/S^6) require more than three dimensions, but I think this will always be because they aren't fundamental. Density is a ratio, mass per volume which is why we get S^6.

I must admit I'm not exactly sure what we have here. It could be of huge importance, or not. I suspect the former, but I'll leave it to you to judge.

Time / Space	F: 0D Space	A: 1D	W: 2D	E: 3D
F: 0D Time	S = Space, Charge 1/S = Power		S^2 = Area $1/S^2$ = ?	S^3 = Volume $1/S^3$ = Luminance
A: 1D	T = Time, Change 1/T = Frequency	S/T = Velocity, Current T/S = Energy, Moment, Torque T*S = Electric Flux Density	S^2/T = Permittivity T/S^2 = Force, Drag, Voltage S^2*T = ?	S^3/T = Compliance, Capacitance T/S^3 = Stiffness, Electric Field Strength S^3*T = ?
W: 2D	T^2 = Change of Change $1/T^2$ = ?	S/T^2 = Acceleration T^2/S = Angular Momentum T^2*S = ?	S^2/T^2 = Conductivity T^2/S^2 = Momentum, Impulse.	S^3/T^2 = Conductance T^2/S^3 = Resistance S^3/T^2 = ?

			Resitivity $S^2 \cdot T^2 = ?$	
E: 3D	$T^3 = \ldots$ of Change $1/T^3 = ?$	$S/T^3 =$ Jerk $T^3/S =$ Inertia $T^3 \cdot S = ?$	$S^2/T^3 = ?$ $T^3/S^2 = ?$ $S^2 \cdot T^3 = ?$	$S^3/T^3 =$ Magnetic Reluctance $T^3/S^3 =$ Mass, Inductance $S^3/T^3 = ?$

Chapter 9

A New Theory of Colour

Introduction

In this chapter, we'll apply the UP to the phenomenon of colour as a test and an example of how the principles are used.

There is an implicit assumption that we should be able to discover an absolute relationship between colours and the entities of the UP. *I.e.* there should be a "right" colour for each, or at least for the 4E / 7P. I had no preconceived ideas of what to expect from the investigation, apart from that.

I chose the question of colour as the example to use here because:
- it should be relatively simple
- I wanted to know the colour-mapping for aesthetic purposes, so I can give the elements colours when I'm writing about them.

I wasn't looking for, or expecting to find, anything controversial, but of course I did.

The investigation led to evidence that the trichromatic theory of colour vision (how we see colours) may not be correct: It suggests there are four primary colours, and colour-detection in the retina works in a radically different way than is currently thought.

I also seem to have stumbled upon evidence that Newton was wrong about how the prism works. I really didn't want to be challenging the giants of physics in this book, but I suppose it was unavoidable.

Applying The Theory

Much of the work of any science is in finding correspondences and categorising phenomena. However, working with the UP is the opposite of material science. Materialist science is "bottom up", from matter to principles. Spiritual science is "top down" from principles to matter.

To understand light and colour from the materialist perspective one must collect data about what light does in the physical world, and then try to figure out if there's a pattern. Using the UP we already have the pattern. To make use of it, we analyse the concept of colour, discover how it fits into the plan, and that should tell us how it works.

The UP provides a basic hierarchy of archetypes into which everything must somehow fit, but it's not always easy to figure out where. We must discover the natural hierarchy behind the concepts, where the dependencies are, and where the one-to-many relationships point.

The phenomenon of colour is dependent on light. The concept of "light" must exist before "colour" can exist. Therefore, they are on different levels of the hierarchy.

The concepts of light and shadow are defined at the level of duality. Yang is light and Yin is shadow. This is basic stuff; we can take these concepts as already defined and understood. If the level of duality (Air) is light/shadow, then colour must occur at the next level, the four elements (Water), where "all the work gets done".

The phenomenon of colour is complicated by the fact that there are, in practical use, two different "colour-models", additive and subtractive, a duality. We use RGB for making colours with light, which is called "additive colour", and CMY (cyan, magenta and yellow) for painting or printing which is "subtractive". However, this isn't a problem. Duality is always our friend.

After some investigation, we find that both systems can be described by a system of four primaries which correspond to the four elements. We can make both the RGB and CMY models from the 4E.

A New Theory of Colour Vision

At the start of this investigation, I wasn't really expecting to find anything of great significance beyond a colour-mapping. As we use the three primary colours of RGB, I wasn't even sure how to reconcile the three into the system at all, so my expectations were low.

After working through it though, again I'm amazed by the explanatory ability of the UP, and how it provides viable answers to questions which have been pondered for years, such as how the eye handles colour.

Without the structure and discipline the UP imposes on any analysis, science is much more difficult. If we instead attempt to understand phenomena qualitatively and in terms of the UP, we find they give up their secrets far more readily.

The theory proposes the following:

1. There are four primary colours required to describe both the additive and subtractive colour models.

2. The sensitivity responses of the cones in the human retina appear to be adapted to view colour according to this four-primary model.

Some interesting points emerge:

1. Of the three "primary colours" RGB, two are actually mixtures: green and blue, but it doesn't actually matter in practical terms.
2. Goethe was effectively right about colour being a "mixture of dark and light", but perhaps not in the way imagined.
3. Newton was wrong about how the prism separates the colours.

Colour Vision

Vision should be a relatively simple sense to understand in principle, as more Yang phenomena are simpler.

The sense of vision is a symbol *aspect of the Fire element (Yang*Yang).

	State	External Sense	Medium
Fire	Plasma / Electricity	Vision	Light, Aether
Air	Gas	Hearing	Sound, Air
Water	Liquid	Smell / Taste	Water (Solutions)
Earth	Solid	Touch / Pain etc...	Matter

While it may be simple in relation to other functions, the human visual-processing system as a whole is quite complex. The (theorised / assumed) subconscious, computer-like part of the mind which turns the raw-data collected from the sense organs into our experience is doing considerable work.

It appears, from looking at certain colour-based illusions, that our visual-system takes the entire visual field into account when processing and attempts to colour-correct for lighting conditions. It tries to provide us with a consistent representation of the colours of objects, no matter what the source of light is.

In the image linked below the squares A and B are the same shade. Our visual processing algorithm can cope with the shadow cast by the green cylinder, and we see a consistently chequered board.

https://en.wikipedia.org/wiki/Checker_shadow_illusion

The visual system must consist of several levels of processing, but we're not trying to attempt to explain the entire thing, only it's beginning, the perception of colour.

More illusions:

https://www.thedesignwork.com/65-amazing-optical-illusion-pictures/

The eye needs to be able to distinguish between situations where it should interpret the colour data as RGB *additive, and others where it needs to use CMY* subtractive. It sees both direct and reflected light in the real world, so it needs to be able to handle both models. The way it does this is with a palette of four primaries.

How Colour Is Made

The "universal form" of creation is FAWE, so colour must be made in that order.

Given the properties of the archetypes, the process of creating colour can only follow the pattern below. We start out with just "light" and end up with seven *eight* infinite colours.

7P: Creation of colours from white light	
Fire	Undifferentiated light (white) Fire is One, undivided, unmeasured / infinite
V	Reflection / Negation (Undifferentiated light is reflected into ...)
Air	White + Black, Light + Shadow Air is Two, choice, difference, distance
H	Alternation (Light and shadow are alternated into ...)
Water	4 Primary colours Must be where the primary colours are made.
X	Mixing (Four primaries are mixed into ...)
Earth	Many Mixed Colours Earth is matter, the final product, the result, the object of desire.

The UP is saying that colour must be made by "alternating light and shadow". This idea doesn't immediately seem to agree with the idea of colour being caused by different frequencies of light, but it's just a question of interpretation.

The alternation is not in the time domain, like switching the light on and off sequentially, but in the frequency domain. White light is the whole spectrum, if we create "shadows" in parts of the spectrum, then we get colours. *I.e.* if we remove bands of frequency from white light, we get colour.

It's interesting that the word "shade" is a synonym of "colour". The idea that colour is "made of shadow" must have been part of mainstream thought when the word came into use.

Note, the table above only deals with the case where colours are produced from white light. Matter produces coloured light of specific frequencies, and it's a mixture of many of these that produces white light. That is a different process and would need its own analysis.

The above predicts there are four primary colours, but it doesn't tell us which colour corresponds to which element. To find that we must go another route.

Darkness From Light

The RGB colour model that we commonly use is additive, it works by adding light to darkness. Without any picture on it, your TV screen is black. However, according to the template, the universe doesn't begin with darkness, it begins with light. The archetypes work as follows:

- Yang is light, Yin is shadow.
- Yang always comes first.

Also:
- Fire is the most Yang element, and it is an archetype of light.
- Fire is also "the beginning" because all creation begins with desire.

The archetypes say that everything begins in light, not in darkness, and that darkness follows, so this begs the question: how can we create darkness from light?

It might seem to be a paradox, but it isn't. There are two physical mechanisms that can do it: interference and absorption. For light to be "absorbed" though, matter would have to exist to "soak it up", and at the beginning of the universe there is no matter.

The process of interference, however, can negate light with more light. You can have two light waves that exactly cancel each other out leaving (the appearance of) darkness.

Interference occurs via "reflection". The two interfering "waves of light are an exact reflection / opposite of each other, and it occurs in nature via (partial) reflection. So, the mechanism of "reflection" fits perfectly here.

This means it's possible to have "darkness that is made of light", but you can't have the opposite: "light made of darkness".

You can have Yang (light) without Yin (shadow), but you can't have Yin without Yang.

The archetypes suggest that the universe is made solely from "light", and shadow is an "illusion" because it's also made of light. (This archetype doesn't only apply to physical light of course.)

- Light is Yang: one direct light.
- Shadow is Yin: an effect of many (two) conflicting, indirect lights.

(Note, reflection and absorption may be a duality.)

Additive vs Subtractive Colour

Light can have a colour. Matter can also have a colour. When we look at an object, the colours we see are a function of the colour of the object and the colour of the light it is illuminated by.

RGB: Red/Green/Blue - Is the colour-model we use for "painting with light".
CMY: Cyan/Magenta/Yellow - Is the colour-model we use for "painting with matter".

These two systems obviously form a complementary duality, so how do we classify it?

- RGB are the "dark" *Yin colours, CMY are the "light"* Yang colours.
- RGB starts with a dark background. The colours must be dark to allow the reproduction of dark tones.
- CMY starts with a light background. The colours must be light to allow the reproduction of light tones.

How do these classify?

In the final analysis we find the Yang colours are Cyan and Yellow, and the Yin ones are Red and Violet However, RGB is used for light and CMY with matter, which suggests the opposite.

So, how do we interpret this? Which model is Yang?

The "Yin and Yang paths"

There's quite a lot of information embedded in these concepts *colour-models, and it's not clear exactly how they should be framed* defined. They are a duality, but seem symmetrical, which implies they are 2D, belonging in the Water category, like magnetism. They may be the "north and south poles" of colour.

For now, I suggest we can frame these colour models as the "Yin and Yang paths". *I.e.* they tell us how to get from one to the other.

- Yang is light. To get colour we must add matter / darkness.
- Yin is dark. To get colour we must add light.

I suppose these models symbolically define what we must do to Yin or Yang to invert them.

	Yin path - Dark to light **Additive, Light**	**Yang path + Light to dark** **Subtractive, Matter**
Model	RGB: Red - Green – Blue Dark colours	CMY: Cyan - Magenta – Yellow Light colours
Direction	From dark to light (Yin to Yang) Have no light, need light to make colour. - "Earth to Heaven"	From light to dark (Yang to Yin) Have light, need matter to make colour. - "Heaven to Earth"

The RGB Colour Wheel

I think the following image shows the best visualisation of the colour-system for our purposes. Note the six big triangles are the main colours from the two models. Each colour in the wheel can be made as a mixture of the two either side, *e.g.* red can be made as a mixture of magenta and yellow.

In this diagram, RGB forms a triangle pointing upwards, from Earth to Heaven, appropriately. CMY is the reverse, as Yang's direction is towards Earth.

The two opposed triangles form the symbol known as the "hexagram", sometimes linked to the Hermetic "as above so below", also the "Star of David". I don't know if there is an actual link there, but it's worth noting. It's possible the symbol originated in Alchemy.

Each colour has a dual opposite, directly across the wheel. The opposite of red is cyan, for example. When added together they will make either white or black, depending on whether light or paint is being used. So, duality re-emerges in the relationships between individual colours, which is interesting to note.

```
https://commons.wikimedia.org/wiki/File:RBG_color_wheel.svg
```

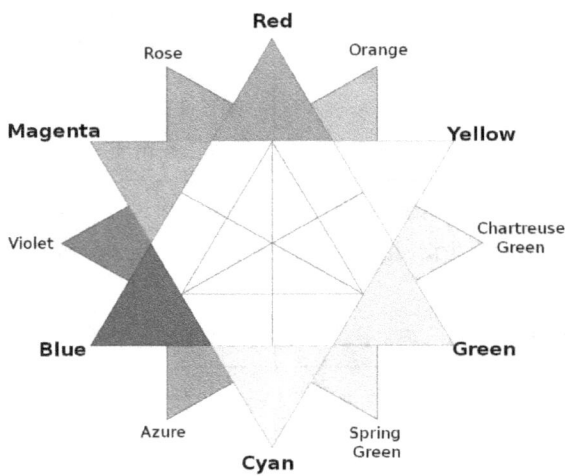

Finding Red's Complement

It's quite easy to demonstrate that red's complement is cyan. You can see it for yourself.

1. Go to the colour wheel diagram above.
2. Stare at the red triangle for 5-10 seconds.
3. Close your eyes, relax, and you should see a light blue triangle as an "after-image".
4. Alternatively, shift your gaze to the white background, and the blue after-image should appear.

With eyes closed I find the opposing cyan is more like the "azure" blue, but on a white background it's very close to the cyan as depicted. What is certain is the complement isn't green (as some have suggested).

You can go around the wheel and repeat this with the other colours to find their opposite, and they match well, although I do find some slight discrepancies. I'm not sure why.

Metamerism

The phenomenon of metamerism allows RGB displays to reproduce millions of visible colours with only three different primary colours / frequencies.

We perceive mixtures of colours as other colours. We perceive green + red light as yellow, for example. We don't have the ability to distinguish between a monochromatic yellow light and a mixture of red and green.

There are many ways of creating the "same" colour (appears to be the same to our eyes). A single shade can be made by many mixtures of other shades. This is called "metamerism".

The word "metamer" means "beyond the parts", so it refers to something that's "more than the sum of its parts". As red + green creates something new (yellow) which is "more than" just a red-green, it's new entity.

In colorimetry, metamerism is a perceived matching of colors with different (nonmatching) spectral power distributions. Colors that match this way are called metamers.

https://en.wikipedia.org/wiki/Metamerism_(color)

The human eye exhibits a property called metamerism, which means that although it has sensors only for red, green and blue, it produces a signal for yellow for example when simultaneously illuminated by green and red light. Similarly, red and blue together yield magenta and by trimming one intensity or another, you can get purple out of the combination.

Because of this property, the human eye can be fooled into perceiving the full spectrum of colors in nature with only an R/G/B source.

https://physics.stackexchange.com/questions/433119/why-does-purple-appear-in-a-rainbow

Video: "What's the Difference Between Yellow and Yellow?"

https://www.youtube.com/watch?v=F8ev_K3kzg8

Article: "What is Yellow?"

https://pscolour.eu/English/whatyellow.htm

This tells us that any individual colour "comes in two forms":

- Yang: monochromatic coloured light (one single colour / archetype)
- Yin: a function of the spectrum as a whole (many possible mixtures / symbols)

One Archetype, Many Symbols

We could change the frequencies of the three "primary" colours in our TV set, but it would still be able to give us the same colours as before if we mixed them differently. There are an infinite number of ways to create an individual colour by mixing other colours.

This tells us that the principle of colour is more fundamental that the physical processes that convey it. If there's one colour, but many methods of making it, the colour is Yang to the method. The colours are archetypes, the expression is the symbol.

This tells us that colours can appear in other contexts, not just as light, but as other things. In fact, everything must have a colour. Everything is made of the four elements, and they have colours.

Given the fastidious efficiency of Nature, we would expect the archetypes of colour to be re-used in multiple other contexts. It would make sense to consider phenomena such as music, or scent, to be analogous to colour and vision in some way.

Lower levels of the UP are made of the higher ones, so this suggests, among other things, that:

- **all qualia can be represented as 2D coloured images (or a pair of 2D images, producing 3D).**
- **all senses are a form of vision**

The UP suggests an underlying commonality between the senses which would explain "synesthesia".

https://en.wikipedia.org/wiki/Synesthesia

(Note, all qualia / information can ultimately be expressed as a "wavy line", a 1D string.)

Current Theory

A brief discussion of the basics of current theory is necessary.

The two most influential figures in the study of colour are Sir Isaac Newton and Johann Wolfgang von Goethe. Some of their discoveries are relevant here, so I'll summarise.

Sir Isaac Newton

We can't discuss colour without mentioning Sir Isaac Newton, whose description of how light is refracted, and colours are produced in the prism has been the scientific model for about 350 years. It was Newton who proposed that white light consists of a mixture of many frequencies which corresponded to different colours, and these could be separated with a prism via the principle of "dispersion", a form of "refraction".

refraction
- the change in direction of a propagating wave, such as light or sound, in passing from one medium to another in which it has a different velocity

dispersion
- the phenomenon in which the phase velocity of a wave depends on its frequency

In optics, one important and familiar consequence of dispersion is the change in the angle of refraction of different colors of light, as seen in the spectrum produced by a dispersive prism and in chromatic aberration of lenses.

`https://en.wikipedia.org/wiki/Dispersion_(optics)`

The principle of refraction doesn't just work with light waves, it applies to all kinds of waves. For example, water waves are refracted / turned by shallows because they are caused to slow down by them.

In 1666, Newton observed that the spectrum of colours exiting a prism ... is oblong, even when the light ray entering the prism is circular, which is to say, the prism refracts different colours by different angles. This led him to conclude that colour is a property intrinsic to light – a point which had, until then, been a matter of debate.

`https://en.wikipedia.org/wiki/Isaac_Newton#Optics`

Newton's theories on light and colour have stood the test of time, but it may be necessary to challenge them nonetheless. He did make a mistake which it seems has never been noticed; I'll explain shortly.

Johann Wolfgang von Goethe

Newton's focus was to describe the phenomenon mathematically, to do quantitative science, whereas Goethe viewed colour more qualitatively and was arguably the better observer. Goethe was a critic of Newton's work on optics, believing that white light was not a mixture but was "undivided" and homogenous.

For Goethe, light is "the simplest most undivided most homogenous being that we know. Confronting it is the darkness"

homogenous
- of the same kind or nature; unvarying; unmixed

(It's possible Goethe intuited the existence of the Yang form of light, as discussed in the section on physics, which sort-of fits this description.)

Goethe saw the phenomenon of colour as emerging from a mixture of dark and light, although he didn't really propose a mechanism for how this happens. Pehr Sällström (see below) has, however, done that by correlating it to metamerism, and "holes" in the spectrum.

Unlike his contemporaries, Goethe did not see darkness as an absence of light, but rather as polar to and interacting with light.

https://en.wikipedia.org/wiki/Theory_of_Colours

Goethe had an interesting, dualistic view of light/dark, and he thought they could interact with each other. Perhaps he intuited the structure of matter and the relationship between the black-hole and light. That is a form of "darkness" which does indeed interact directly with light.

Anyway, it's Goethe's excellent observations which are helpful here. He showed that light refracted by a prism emerges as four colours, in two sets of two. They then merge to create a new colour in the middle. This observation is crucial, and we'll come back to it.

Pehr Sallstrom on Colour

Pehr Sällström's website is an invaluable resource for understanding colour and for building a more scientific understanding of Goethe's views. He provides some solid theory where Goethe didn't. He explains how colour works with much more depth than most can offer.

https://pscolour.eu/

I would recommend this presentation which explains how colour should be thought of in terms of the whole spectrum, not just as single frequencies.

https://pscolour.eu/Basel/ColourPhysics.htm

As the phenomenon of metamerism demonstrates, the colours we see do not necessarily correspond to certain frequencies. The eye is sometimes said to "average" the wavelengths it receives to arrive at a decision on which colour to "display", but it's not a simple average, that would be too crude a mechanism.

https://pscolour.eu/English/whatyellow.htm

Mr Sallstrom has developed some software which allows you to see what colour you get given a certain spectrum of light, and it shows that there are many different spectra which can give rise to the same perceived colour.

https://pscolour.eu/English/visualspd.htm

Trichromatic Theory

The Young Helmholtz theory / trichromatic theory was first proposed in 1802.

In 1802, Young postulated the existence of three types of photoreceptors (now known as cone cells) in the eye, with different but overlapping response to different wavelengths of visible light.

https://en.wikipedia.org/wiki/Young%E2%80%93Helmholtz_theory

It was expected the cells' sensitivities should be evenly spaced out across the spectrum, corresponding more or less to the RGB model.

Hermann von Helmholtz developed the theory further in 1850: that the three types of cone photoreceptors could be classified as short-preferring (violet), middle-preferring (green), and long-preferring (red)

Their theory is now considered to have been proven correct, except for the predicted sensitivities.

The existence of cells sensitive to three different wavelength ranges (most sensitive to yellowish green, cyanish-green, and blue – not red, green and blue) was first shown in 1956 by Gunnar Svaetichin

The sensitivities of the three cone types to the visible spectrum has been measured, but it doesn't appear to be optimally "tuned" for RGB. The reasons for this aren't clear in current theory. While trichromatic theory has been shown to be right about the three types of cone, their mode of operation differs from the prediction. It seems they aren't "looking for" RGB, but something else.

Another odd feature is that the blue cone is much less sensitive and numerous than the other two. This again is not understood. The four-primary system described here can explain both apparent anomalies.

Observation of Prisms

Goethe's detailed observations of prisms were most helpful in formulating this theory. When searching for diagrams of how prisms separate light (dispersion), the results tend to look something like this.

https://commons.wikimedia.org/wiki/File:Dispersions.png

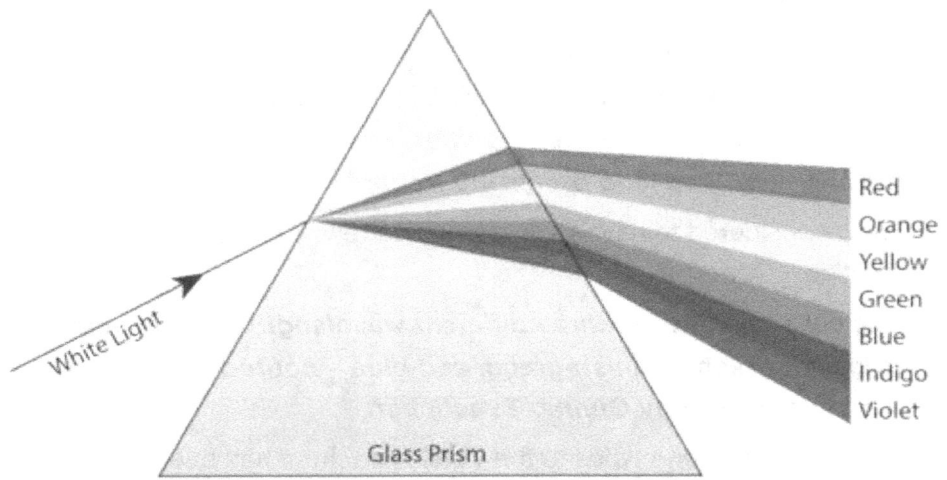

This is a poor representation of what happens in real life.

The following, far more accurate, diagram from Goethe shows how light emerges from the prism, with colour appearing as four bands. This is more like what we see.

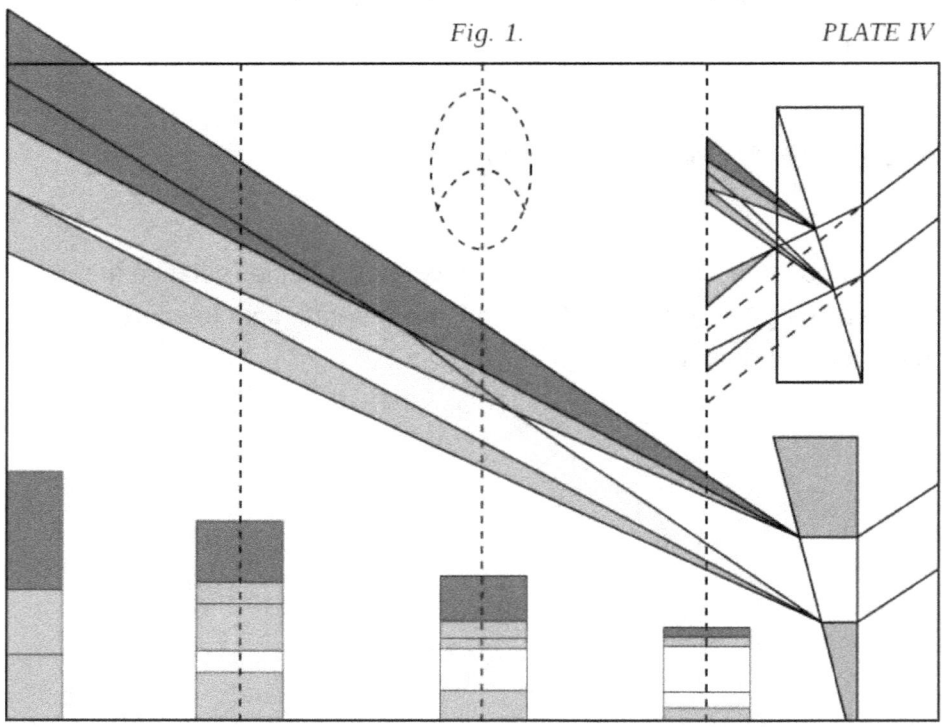

https://en.wikipedia.org/wiki/File:Goethe-LightSpectrum.svg

We don't see colours inside the prism and at the surface where the light emerges there is no colour. The colour only develops at distance from the prism, apparently originating at its surface. When the colours first emerge, there are only four of them, and they appear as homogenous bands. Green is formed when yellow and cyan start to overlap.

In the above drawing the lines between the colour bands are very sharp. In real life the edges aren't as sharp, but the bands are still clearly a single block of colour, and not a range.

The best way to see the effect is to look directly through a prism. You can clearly see uniform blocks of colour, although they do mix where they meet.

I propose these four initial colours we see from the prism are the four (true) primary colours our eye is capable of detecting.

Four Colour Bands

I think it true to say that the colour banding we see is not explained by trichromatic theory.

The bands look like blocks of a single colour, but each must, presumably, contain a range of frequencies. The output from a prism should be a continuous spectrum of monochromatic light, it shouldn't be "quantised" into discrete bands, but that is what we see.

Also, we might ask why do neither the yellow nor cyan bands reach green in the image below? There's not even a tinge at the far edges. If the prism is separating out the frequencies, and green is a frequency which we can detect, then why don't we see it here?

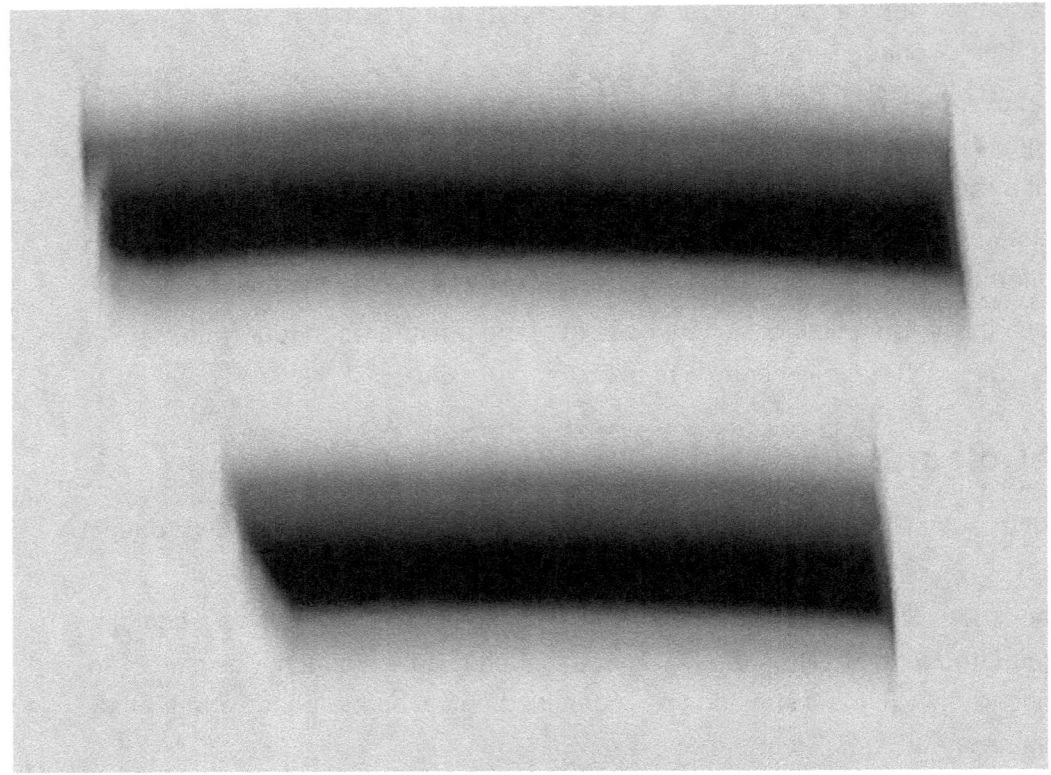

(This photograph shows two pieces of black tape on white paper, viewed through a prism.)

Four primaries

If the eye is receiving a range of frequencies, but can only perceive a single colour, it implies that must be all it can recognise across that range. That would seem to be the most obvious interpretation.

- The prism produces a continuous spectrum of monochromatic light.
- But we see discrete bands of colour instead.
- Therefore, we cannot see a continuous spectrum.
- Therefore, some colours are "primaries", and all others only exist as "mixtures".

The evidence suggests the eye detects colour in four frequency bands, corresponding to red, yellow, cyan, and violet, as we see in the main bands of colour from a prism.

It means we must divide colours into a set of primaries, and a set of mixtures, as the UP predicted.

Mixture colours

Where two primary colours combine, the mixture colours appear. They are:

Red + yellow = orange
Violet + cyan = blue
Yellow + cyan = green
Red + violet = magenta (purple)

Wide "seasons", narrow "events"

The UP suggests that the eye's perception of a monochromatic light spectrum should take roughly the same form as the layout of the yearly seasons. Both are waves / cyclic phenomena, so we should correlate them.

- The four primary colours correspond to the seasons
- the four "events" (solstice / equinox) correspond to the mixture colours.

Therefore, the mixture colours (orange, blue, green, magenta) will occupy relatively narrow frequency bands, primary colours will appear as wide bands.

Four-Primary Colour Theory

We'll call this model "Four-Primary Theory" (4PT).

The claim here is that the eye's primary colours are: Yellow, Cyan, Red and Violet. YCRV.

A caveat

The biological reality of colour vision is extremely complex. There is a wide variety of visual systems in the animal kingdom, and even within the human species there is diversity in colour recognition. Having said that, 4PT does seem to describe normal human colour vision accurately.

I assume this theory describes the fundamental template / archetype of colour vision in general, and that human colour vision essential conforms to that archetype (more or less) perfectly, whereas other systems use variations on it.

I suggest that the purpose of biological life is to "discover the basic archetypes, and then build with them", and this corresponds to "evolution". Life starts out as Yin, not knowing. Matter must discover the rules it is subject to, and then use them to its advantage. So, we wouldn't expect all living things to have the same visual system any more than we would expect all people to have the same opinions.

4PT describes the underlying archetype of colour vision, it is not suggesting all creatures must employ it in exactly this form, but it is suggesting that normal human colour vision does.

The qualities of colour

The UP is a qualitative theory. It claims that reality is made of qualities and that quantities are a secondary phenomenon. Our senses should be considered as being primarily concerned with qualities.

The eye is looking for qualities not quantities.

The most fundamental qualities (properties) that colour can have should be a complementary duality, and we should be able to fully define it as a function of those qualities. So, what are the basic qualities of colour?

- The proposed complementary parent-duality is **temperature / brightness**.
- The opposing child-dualities are **hot / cold** and **light / dark**

Light and Dark Colours

I want to draw an objective distinction between the "light" and "dark" colours because I suggest this is a fundamental property of colour.

We can think of colours in terms of their RGB content. In this model we're not thinking of colours as being at specific wavelengths but as a mixture of three different colours, like the pixels on a TV. The RGB model splits the colour spectrum up into three sections, and then we can quantify how much of each is in each colour.

Each of the colours in the wheel above can be made of mixtures of these three, and you can add and subtract colours using this system quite easily. There are six primary colours in the wheel, and they correspond to the six possible on/off combinations of RGB.

Colour	Red	Green	Blue
Red	On		
Green		On	
Blue			On
Yellow	On	On	
Cyan		On	On
Magenta	On		On

If we think of each primary colour as a LED bulb, then the top three colours, RGB, are "dark" colours because they only have one bulb lit. The "light" colours have two bulbs lit, so they're brighter. This objective distinction between light and dark colours is important.

RGB are "dark colours". CMY are "light colours".

Temperature and Colour

Colours have a "temperature", we perceive them as warm, cold, or neutral. Red *yellow is perceived as warm, and blue* cyan is cold. The argument here is that "temperature" is the other fundamental property of colour. There are "warm" and "cold" colours.

In terms of our perception, we could view the colour / temperature scale like this.

Blue, Cyan	Green	Red, Yellow
Cold Ice, Snow, Water	Mild Green Plants	Hot Desert, The Sun, Dryness

In some cases, we can determine an item's temperature directly from its colour, and we can talk of colour-temperature. When heating a piece of metal, it's colour will go from dull-red to yellow, white then white-blue at its hottest. This is called "black body radiation".

`https://en.wikipedia.org/wiki/Black-body_radiation`

In black-body radiation the temperature assignment to colours is reversed with red being the "coolest" and blue being the hottest temperature. Relative to us though, anything heated enough to glow red is still very hot, of course. We don't usually see things which are blue-hot. In metallurgy the temperatures used go no further than white-hot, 1500C.

`https://en.wikipedia.org/wiki/Red_heat`

The colours exhibited by black-body radiation do not correspond to the system described here, and it lacks some colours. I'm not sure how to relate these phenomena at this stage. Matter is complicated.

Conceptually however, from a human perspective, temperature is generally related to colour as described in the table above. Red is warm, blue is cold.

An objective classification

So, the two fundamental properties of colour are: light *dark, and hot* cold. This is equivalent to brightness *temperature which is a complementary duality. This is effectively linking the concept of "activity" to that of "outside* inside". Temperature is "activity inside", brightness is "activity outside".

I think it's true to say these are all objective qualities, they aren't subjective judgements.

While the "light/dark" property is clearly objective, the "warmth" or "coldness" of colours as perceived by us is arguably less so. The proposal here, however, is that "red" is objectively defined as "warm" within the UP.

- The 1D quality of "warmth" must have existed before the 2D colour "red" could be created.
- The internal definition of red (in the UP) is "the warm dark colour".

Notably, red is the colour of light which transmits physical warmth. So, it doesn't just look warm.

The Four Primary Colours

We can plug these qualities into the table of the four elements, and it should give us the colour assignments.

"Temperature" is the leading / Yang property because it drives the other property of "brightness". The hotter something is the brighter it glows. I propose that these are the most fundamental properties of colour, they define the colour, not the other way around.

Note: there is some ambiguity in the table below between blue and violet to compare it to RGB, but the principle is still valid. Violet / blue is the dark, cold colour, cyan is the light cold colour.

If we fit the colours to the elements according to the rules, we get the following correspondence.

Element	Colour	R	G	B	Light/Dark	Warm/Cold
F ++	Yellow	+	+		Light (+)	Warm (+)
A +-	Cyan		+	+	Light (+)	Cold (-)
W -+	Red	+			Dark (-)	Warm (+)
E --	Violet / Blue			+	Dark (-)	Cold (-)

We did not fit the colours to the elements by any subjective measures, but by objective qualities. Fire isn't Yellow because it's "like the sun" for example, but because the binary bits of qualitative data describing its properties define it as such. There's no ambiguity in this arrangement.

Two bits of qualitative data

This is the proposal:

There are four primary colours which convey two binary bits of qualitative information: "lightness" and "warmth".

Yellow is the colour which conveys the information of "light and warm", red is "dark and warm". Cyan is "light and cold". Violet is "dark and cold". This may seem a peculiar way to classify colours, but this is how the universe itself defines them.

As we shall see, this arrangement then goes on to predict the order of colours in the spectrum. Which is an amazing result. It means we will have effectively derived the colour spectrum from first principles.

The UP makes strong predictions about the nature of colour. It says there should be four primaries and describes how they are constructed. It explains what qualities the eye should be looking for, and how it communicates that data to the brain.

The next question is, does any evidence support it?

Six Colours from Four

The first item of evidence I would offer, is its explanatory power.

We might ask, why would the eye need to detect four primary colours?

Potentially, it provides a simple mechanism that allows the eye to handle both the RGB/CMY colour models. It allows the eye to be able to differentiate between a light source and reflected light from an object, and correctly interpret their colours. I assume RGB alone would not contain enough information to do this.

This is how it works.

Element	Colour	Subtractive, CMY	Additive, RGB
F	Yellow	Yellow	Yellow + Cyan = Green
A	Cyan	Cyan	
W	Red	Red + Blue = Magenta	Red
E	Blue (Violet)		Blue

For the RGB model, we have R and B, and Y + C is Green.
For the CMY model, we have Y and C, and R + B is Magenta.

The symmetry of these two-colour models is a work of art in itself.

Note, according to the UP, anything that is symmetrical must be the Water element, hence circular. The circle is the origin of symmetry in the universe.

That covers the basic qualitative colour model.

How The Eye Sees Colour

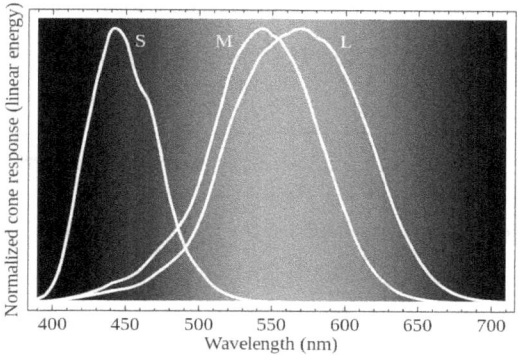

https://en.wikipedia.org/wiki/File:Cone-fundamentals-with-srgb-spectrum.svg

The graph shows the measured sensitivity of the three different cone types to the visible spectrum. It doesn't convey some of the important features and a better image can be found at the link below, but can't be included here.

https://www.simplypsychology.org/what-is-the-trichromatic-theory-of-color-vision.html

The features which can't be explained by trichromatic theory are:

- The red cones' sensitivity is not as far toward the red end of the spectrum as we would expect, it's almost overlapping with the green.

- The blue cone's sensitivity is lower the other two types (not shown above).

Another important feature not shown above is that the red becomes more sensitive than green in the violet range. This is crucial in explaining how violet is detected.

The "Four-Mode Pixel" Model (4MP)

Trichromatic theory imagines the eye works something like a RGB video camera, *i.e.* each cone acts independently to return a continuous intensity value. The resulting colour is then derived from the three analogue RGB signals.

The UP suggests that the eye operates in an entirely different way. The colours are not derived directly from the cones, but indirectly, via the relationships between the three types. Also, they return a digital, not an analogue signal.

- Individual cones do not return a continuous / analogue colour signal.
- Cones always work as part of a "pixel".
- Colour is derived as a function of the relationship between the cones' signals.
- There are four relationships, hence four primary colours.

Instead of a "three-colour" model, the UP suggests a "four-mode" colour model. Here's a list of the main differences between the theories. They are almost opposites.

Trichromatic Theory	4MP Theory
Individual cones produce a continuous signal. Cones send data directly to the next level.	Cones do not operate individually, only as a pixel. Cones send data indirectly, via "consensus".
Output: RGB - analogue	Output: Colour data (YCRV) - digital Intensity data - analogue
Colour is a function of the frequency of light. Cones detect quantities.	Colour is a function of the relationships between qualities Cones detect qualities.
Three primary colours	Four primary colours / modes
Colour does not exist in reality. It is "invented" by the brain.	Colour does exist "out there" in reality. It is "detected" by the eye.

Mapping to qualities

It does seem somewhat unintuitive to propose four colours from three cone types. It's probably not an idea one would come up with without the UP as a guide and I was surprised it worked out so well.

So, how do we correlate the cones' sensitivities to the four primary colours?

Conveniently, it turns out that we can map the two basic properties of colour described in our analysis above directly to the relationships between the cones in the above graph.

The cones in the eye operate as a duality.

- The red and green cones work together to determine if the colour is "light" or not.
- The blue cones determine if the colour is "cold".

The mapping is as follows.

Light: The light colours correspond to "there is more green than red"
Dark: The dark colours correspond to "there is more red than green"
Warm: The warm colours correspond to "there is no blue"
Cold: The cold colours correspond to "there is some blue"

So, this is how the model predicts the eye distinguishes between the four primary colours.

	Colour	Light colour? = Green > Red	Warm colour? = No Blue
F ++	Yellow	Yes (+)	Yes (+)
A +-	Cyan	Yes (+)	No (-)
W -+	Red	No (-)	Yes (+)
E --	Violet	No (-)	No (-)

Note that the light *dark mapping is a comparison between the red and green cones, whereas the* warm *cold mapping is a simple binary on* off. *A comparison is complex, but a binary on*off *is simple. This is a Yin* Yang *relationship, blue is one* simple, green/red is many / complex.

Each primary colour is defined as two binary bits of qualitative data, so it's a 2-bit (2D) colour model.

The "Pixel"

For this system to work, the visual system must be treating groups of the three-cone types as a "pixel". In this way it is similar to the "digital camera model". The colour returned by a pixel is a function of the signal from all three cone types within it.

A "pixel" in this case is loosely defined and could be implemented in "software" or "hardware", *i.e.* in the higher-processing layers or in the retina itself. A pixel must consist of three individual cones at minimum but could include more.

pixel (eye, my definition)
- logical grouping of the three different types of cones into a single unit
- capable of returning a 2-bit, discrete colour value, one of Y/C/R/V
- also returning a continuous intensity value from the "rods"

This theory suggests luminance / intensity data is probably only provided by the rod-cells, so a pixel should be thought of as including rods as well. Each pixel thus returns a (2-bit, discrete) primary colour from the cones, and a (continuous) intensity value from the rods. One discrete and one continuous data value, forming another nice complete duality.

Pixels return a duality of data:
- a 2-bit digital, qualitative colour signal

- **an analogue quantitative intensity.**

Just to be clear: individual pixels cannot detect mixture colours. The mixtures are detected by combining inputs from multiple adjacent pixels, at the next level of processing "up the chain".

We can classify rods and cones into Yang / Yin respectively, as follows:

Cones: Yin – Colour (Shade)	Rods: Yang - Light
Complex – Only work as part of a pixel	Simple – can work as an individual
Indirect – Output is processed, the result of an algorithm	Output is unprocessed and direct
Discrete – output is 2D, 2-bit digital	Output is 1D continuous.
Derived / secondary Cones are derived / specialised forms of rods.	Primary – Rods came first. The ability to detect light must come before colour.

Mysteries Solved

This arrangement explains the two apparent anomalies with the cone sensitivities.

1. The spectral proximity of the green and red cones.

Trichromatic theory predicts that the responses of the red and green cones should be reasonably well spaced apart in the spectrum, but this is disproven by experiment. Instead, these two types of cone are quite close spectrally. There doesn't seem to be a good reason for this in the current model.

This theory states the red cone is adapted to be used in conjunction with the green. They form a duality, and are not designed to function alone. The pixel is only concerned with the relationship between R and G; it's the comparison between them that determines the colour.

The pixel doesn't need to know the exact wavelength of the red light it receives, it doesn't need to know how red the red is. It just needs to know if there is more red than green. This means the red cone's peak only needs to be a relatively small distance towards the red end of the spectrum from the green. In fact, it's better if the peaks are close in this arrangement, because it gives a sharper division between the two modes.

2. The low sensitivity and number of the blue cones.

Blue cones are less sensitive to light than the other two types, and there are fewer of them. A reason for this doesn't emerge from trichromatic theory but makes sense in the present one.

Blue cones are less sensitive because all they do is detect "some blue". The algorithm doesn't care how much blue there is, they only need return true / false, one binary digit.

A possible reason there are fewer of them is because they don't need to be compared to anything else.

The red and green ones must be compared to each other, and comparisons can be subjective. This suggests a larger sample may be required for error-correction / averaging. As there's no need for averaging with the blue cones, fewer are required.

Flexibility

The system described here might sound very rigid, but it still allows a lot of flexibility.

For example, the theory states the blue cones only provide a simple on/off, but their sensitivity-curve is still a curve, the response isn't completely flat, which suggests there is flexibility built-in, presumably to handle a range of lighting conditions.

There will be natural variations in response between individual cones and pixels. Some of that variation could even be dynamic, imposed by higher levels of the architecture. There could conceivably be feedback and weighting mechanisms that bias how a pixel processes its three components.

Ultimately colours are perceived relative to each other, and in the context of the entire visual field. There are many levels to the process, this is just the first.

Colour blindness

The phenomenon of colour blindness is again rather complex, and I apologise for not attempting to tackle it in any depth here. However, I suspect 4PT can shed some light on it.

For example, red-green colour blindness (RGCB) can be caused by "faults" in either the red or green cones, but the effect on vision is largely the same. This indicates those cones work together, as predicted by 4PT.

It's interesting to note that RGCB equates to being unable to see the colour cyan. People with the condition see it as grey. Green minus cyan would be "dark yellow", which is red. So, perhaps it's really "cyan blindness".

In this system it is likely that simple errors would result in the inability to see a single primary, like cyan.

One insight this theory may be able to provide is a more accurate representation of how the various forms of colour blindness are perceived. *I.e.* does someone with RGCB see reds and greens as red or green, or some other colour?

Well, if it really is "cyan blindness", then what remains would be yellow, red, and violet. They would be looking at the world through "rose-tinted spectacles".

Anyway, this is another topic which we'll have to leave for now.

We have now covered the four primaries in detail. The next thing to consider is the mixtures.

The Spectrum

The list of colours in the table of the 4E above is not in the order that we see colours in nature. Yellow is not the first colour we see in the rainbow, red is. What pattern of the 4E will we find in the rainbow? Will it be a new one?

If we map the list of colours onto the visual spectrum, *i.e.* in the order the colours appear in a rainbow, it looks like this.

	Colour	**Range (approx)**	**Green > Red**	**No Blue**
W -+	Red	> 600nm	No	Yes
F ++	Yellow	600-500nm	Yes	Yes
A +-	Cyan	500-450nm	Yes	No
E --	Violet	< 450nm	No	No

The ordering of the four elements we find is WFAE, the sinewave pattern. This is a great result!

The odds of this pattern emerging by chance is 8%. It could have been any of the 12 possible arrangements of the four elements. There seems to be a real, natural correlation here.

Note that I didn't "fiddle" this result in any way. The light *dark, warm* cold classification is objective, and the way it's correlated to the 4E is the same as in every other example we've covered.

Deriving the spectrum from first principles

We have effectively just derived the order of colours in the spectrum from the first principles of logic!

This achievement is so far beyond what I expected at the outset, it's hard to believe it just happened.

Of course, we didn't know before the beginning of the investigation what the basic dualities of colour were, or that the spectrum took the WFAE form, but if we had, we could have predicted it.

The definition of "theory of everything" we're using is "a theory that all natural phenomena can conceivably be derived from", and this meets that condition, it is evidence that the UP is the right answer.

Octaves of colour

The order of colours in the spectrum matches the order of elements in a sinewave / the seasons. It couldn't really be any more perfect. This table links the seasons with the main stages of life, and to the spectrum.

Element	Season / Event	Stage of Life	Colour
Water	Winter	Childhood	Red
W + F = *	Spring Equinox	Puberty	Orange
Fire	Spring	Young Adult	Yellow
F + A = Voice	Summer Solstice	Parent	Green
Air	Summer	Raising Family	Cyan
A + E = *	Autumn Equinox	Family Have Grown Up	Blue
Earth	Autumn	Old Age	Violet
E + W = Sex	Winter Solstice	Death + Rebirth	Purple / Magenta

* See section on "systole / diastole" below.

As colour is a cyclic/wave phenomenon, following the WFAE pattern, it means it would be possible to have "octaves" of colour like we have octaves in music.

This suggests that animals with four cones may have one which detects the octave.

Base qualia

The colours above are something like our "base qualia". They are the raw components of them.

All senses are a form of vision, and so all qualia can be represented as pictures made of colours.

All pictures are information and can be represented as "holes in a continuum", a 1D string of data.

So, we can see how reality can be constructed from the 1D realm of "properties" up into the 2D realm of waves and colour, and then the third dimension is where it all gets mixed together.

The Eight-Colour Spectrum

As there are seven colours in the rainbow, I initially tried to fit the colours to the system of the seven-principles in the order FAWE, but it didn't work. The list of elements starts with Fire/yellow, and the spectrum starts with red (or violet) so it doesn't match.

As a system of eight entities though, it fits perfectly into the WFAE pattern, and includes the missing magenta, which joins violet to red, and gives us a full circle. The primary colours are the "pure elements", the "seasons". The secondary colours are "equinoxes" and "solstices" and are all mixtures.

There are seven "visible" colours and one "invisible" one which doesn't appear in the rainbow, magenta / purple (which was considered the "colour of royalty"). It corresponds to the Sex principle, and to death and rebirth, and I suppose those things are usually done in private.

There's plenty to discuss concerning the individual assignments, and what they might mean, but that would be a discussion for another day.

The Colours of The Human Body

If we fit the elemental colours over the human body, in the 7P pattern we get the following assignments, with Heart appearing as white. I'm not sure if this is significant or of any practical use.

Principle	Colour

Fire	Yellow
Voice	Green
Air	Cyan
Heart	White
Water	Red
Sex	Purple / Magenta
Earth	Violet

Two Conflicts

I should mention that the system of seven "chakras" that is found in eastern traditions, as it is commonly portrayed, does not match the UP, the order of colours is reversed. (See link for example.)

https://commons.wikimedia.org/wiki/File:ColouredChakraswithDescriptions.jpg

I don't know what the significance of this might be. I would hesitate to suggest the above assignments were "wrong", however there is a conflict here. There is also a physics problem to be resolved.

Light at the blue end of the spectrum is believed to carry more energy, so is arguably more "active", and that would suggest it's Yang to red light. UV radiation can do damage to living organisms, violet seems to be more "violent". Higher frequency light seems more "energetic", but paradoxically lower frequencies are more "warming".

There seems to be a conflict between the size / wavelength of light and the archetypes. High frequency, short wavelength light is thought to have more energy, so it should probably be Yang. It's fast, small, active. Yet this is the blue end of the spectrum.

This, at least, suggests there is a more complex 2D relationship to be described here, but it will have to wait.

Systole / Diastole

In the WFAE cycle, we have both the Voice and Sex principles, but the Heart seems to be missing.

Why don't we have the Heart operator (A+W) in the cycle? It's the core process, it should be there.

We do have two unassigned mixtures though, W+F and A+E, the active and passive groupings.

The answer seems to be that W+F and A+E correspond to the two beats of the Heart, so it actually appears twice. In this case a year is equivalent to one full "pump" of the "planetary heart". So, the Heart is indeed present, and the Spring and Autumn equinoxes correspond to its two beats.

The Winter-Spring equinox corresponds to the systole phase of the heart cycle, the active part, and the Summer-Autumn to diastole, the passive part. It's the charge / discharge cycle in another form.

This is an interesting finding; it shows we can connect the linear FAWE and circular WFAE patterns via the Heart. In the former system it takes the single form A+W, in the latter it appears as a duality.

It suggests we can "square the circle", we can translate between the linear and circular frames of reference via the Heart principle, which makes perfect sense as that is where rotation begins.

	Season	**Life**	**Work / "Decision"**	**Form**
W	Winter	Childhood	Charging (with energy) "Process"	Action, activity, work
W + F		Puberty	**Systole**	
F	Spring	Young Adult	Discharge, "Output"	Will, desire, beginning
F + A = Voice		Parent		
A	Summer	Raising Family	Charged (e.g. with "duty") "Choose" (As parent, you choose)	Law, rules, plans (Parent must follow rules of parenthood)
A + E		Family Grown Up	**Diastole**	

			Discharged (from duty) "Input" - Facts (Wisdom, knowledge)	
E	Autumn	Old Age		Product, result, end
E + W = Sex		Death + Rebirth		

This relationship seems to show a transformation, or a translation between the two systems. It seems to be tying up some "loose ends" I hadn't even become aware of yet. I'm not sure where it leads, but it looks useful.

Conclusion

This short investigation yielded much greater returns than I expected. It was a good "work out" for the theory:

- A solid foundation for understanding the principle of colour, built from objective properties.
- Good evidence of the consistency of UP theory via the results obtained.
- A new theory of colour-vision which provides feasible answers to existing questions.

Also, it has led to a new understanding of how prisms work, and to results that indicate they have been misunderstood for centuries.

A New Theory of The Prism

During the investigation evidence emerged which appears to necessitate a new theory of how the prism works. Two observations, detailed below, essentially disprove the current model.

I suggest we can explain the prism's behaviour without the proposition that different frequencies of light have different "indexes of refraction". Newton's concept of "diverse refrangibility" appears to be redundant.

The proposed model is more parsimonious and provides a more accurate description of the phenomenon, but there are aspects of the prism's behaviour that still need further explanation.

Newton's mistake

It seems Newton made a mistake in his analysis of the prism, and it's one that appears to have passed everyone by. I can't find any mention of the phenomenon.

Newton's apparent mistake was that he didn't look into the prism.

In all analyses of how a prism works, it is said that red exits at the "top" because it's "bent less by refraction". Everyone agrees that red exits at the "top" of the prism (e.g. as shown in the diagram in the section "Four Primary Colour Theory").

We do indeed see this ordering when the light leaving the prism is projected onto a surface. However, when we look into the prism instead, we see the opposite, with blue at the top and red below. Here's a photo I took. The white rectangle is a small piece of white paper on a black background below the frame. (I apologise if this is printed in black and white.)

This observation cannot be explained by current theory.

We see the colours are in the reverse order, with blue at the top. What does this mean?

There are two questions to answer here.

1. Why is blue at the top and not red?
2. Why is the colour order reversed when the light is projected onto a surface?

Dispersion

There is another simple experiment which also can't be explained by current theory.

First a quick overview of what is currently believed to be the cause of the colour separation.

The theory of dispersion states that different frequencies of light are bent / slowed by different amounts when passing through the body of a prism.

dispersion
- "The separation of visible light into its different colors is known as dispersion. ... different frequencies of light waves will bend varying amounts upon passage through a prism. ... The absorption and re-emission process causes the higher frequency (lower wavelength) violet light to travel slower through crown glass than the lower frequency (higher wavelength) red light."

https://www.physicsclassroom.com/class/refrn/Lesson-4/Dispersion-of-Light-by-Prisms

According to this model, the colours are separated by their passage through the prism. The difference in the angles of refraction between the different frequencies are present within the body of the prism and are caused by it. It is sometimes called "bulk dispersion" (bulk = body).

However, the following experiment shows that the colours can be generated solely at the surface of a prism, without any possibility of it happening within its "bulk".

Total Internal Reflection

The image below is of an experiment in "total internal reflection" (TIR). It shows how light going from glass into air above a certain angle is completely reflected back into the glass. The interesting part is, before it reaches that angle, while the light is only partially reflected, the exiting light generates a spectrum.

The video this image is taken from is "Total Internal Reflection" from the "QuantumBoffin" channel on YouTube.

```
https://www.youtube.com/watch?v=NAaHPRsveJk
```

If the image doesn't print well, it shows a spectrum, with red at the top as usual, exiting the flat face of a semi-circular glass prism. This is occurring at an angle just below the angle of total internal reflection, where the light is being partially reflected.

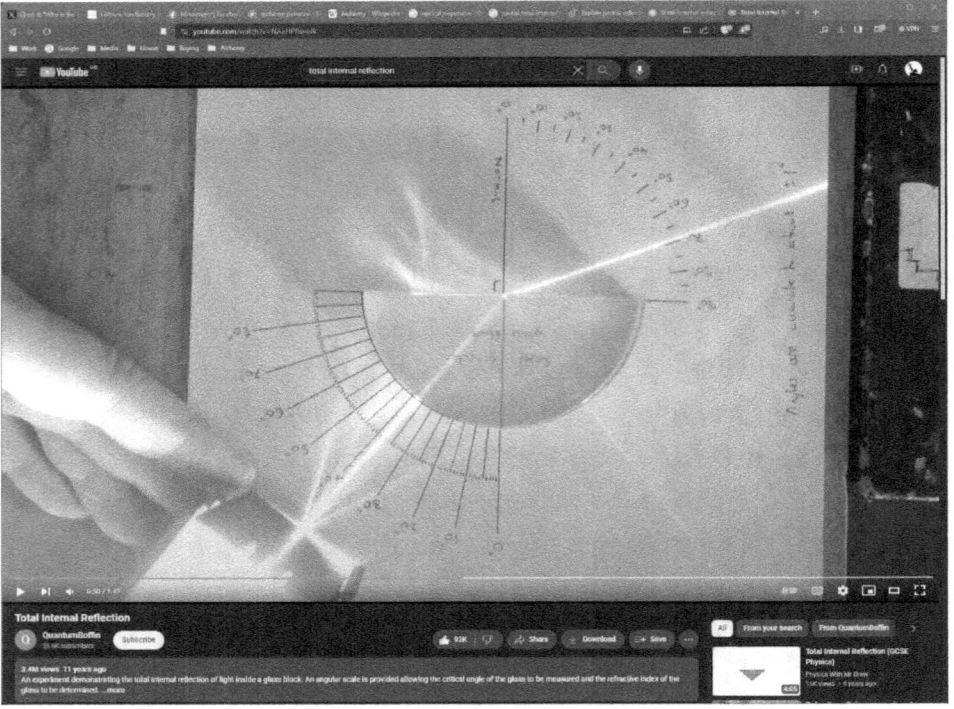

Note that the light is entering the glass prism perpendicular to the glass because it's a semi-circular prism. In this case the colours cannot be separating on entry into the prism or during their passage through its body, because the incident angle is 90 degrees.

Refraction requires the light to enter the prism at an angle to be "bent" and produce a spectrum. If it's perpendicular, there can be no refraction, the light will pass through unaffected. So, in this case, the light must remain "undispersed" until it tries to leave.

In the TIR experiment, the colours can only be occurring on exit of the denser medium, at the surface.

If we can obtain a spectrum without any possibility of dispersion happening within the prism, then the body of the prism is probably not involved in the dispersion effect. There must be another mechanism in action.

There doesn't seem to be any alternative conclusion. This, and the previous experiment, appear to refute Newton's theory. Dispersion does not occur, as is commonly believed, within the body of the prism.

If dispersion can happen at a surface, it probably only happens at surfaces.

If it's possible for colour separation to happen, as in this case, solely on exit of the medium, there's no need to posit any other locations for it to occur. We can explain the phenomenon purely as something that happens when light leaves a denser medium for a lighter one (or vice versa?) at an angle near total internal reflection. We (probably) don't need to think about its entry, or its passage through the medium at all. This makes it much simpler.

How then, are the colours are separated on exit? Why does light form a spectrum when changing media at an angle? If it's not by "bulk dispersion", then what is the mechanism?

Dispersion vs interference

Nature is the most efficient machine ever devised, perhaps conceivable. The proposition that it would have two different mechanisms to create colour seems unlikely. Colours can certainly be created by interference, and the UP seems clear that this is the primary mechanism.

The UP says colour is created by alternating light with darkness. If we consider a diffraction grating, it's easy to see how it conforms to this archetype. It is literally an alternation of opaque lines which block light with transparent lines which allow light to pass. This is probably all we need.

An Interference Model of the Prism

I believe we can explain the prism with just the two principles of (standard) refraction and interference, and the hypothesis of "bulk dispersion" and variable indexes of refraction or (phase) velocities for different frequencies is redundant.

Problem

The current model of "dispersion" in the prism as first proposed by Sir Isaac Newton cannot be correct.

- The first experiment (looking through the prism) shows it does not account for observation. It can't explain why we see blue at the top, it says we should see red.
- The second experiment (TIR) shows that dispersion happens at the prism's surface.

Observation

The second experiment also shows that dispersion only happens when partial reflection is also happening.

That seems to be an elephant-sized clue as to what is really going on.

Solution

The dispersion of colours is due to the surface of the prism acting as a diffraction grating.

- If partial reflection is occurring at the glass / air boundary, that means some of the light is getting blocked.
- Therefore, at the surface, the light leaving the prism is not continuous. It's full of "holes" where some of the light was reflected back into the prism.

This must mean the surface of the prism is acting just like a diffraction grating. Some of the light gets through, some of it gets blocked. We have "alternation" between light and holes. I think it's true to say that:

Partial reflection necessarily means the surface must act as a diffraction grating.

Every "photon" that is reflected from the surface back into the glass represents a reflective "dot" in the diffraction grating. It causes a hole in the outgoing light. When the partial reflection reaches a high enough proportion, the prism produces a clear first-order spectrum, exactly as a diffraction grating does.

Notes:
- In the case of the triangular prism, both the entry and exit surfaces may be acting as diffraction gratings.
- The density (holes/dots per unit area) of the diffraction grating / surface depends on the angle of the incident light.

As the angle approaches TIR, partial reflection increases because the light is more likely to encounter a "dot" / reflection point on the surface. This is equivalent to the density of the bars in a diffraction grating increasing.

Reversed colours

If the prism surface is acting as a diffraction grating, then we would expect to see blue at the top, not red.

The zeroth order would be perpendicular from the prism's surface, and so when we use a prism, we are seeing the first order spectrum. We would expect to see blue nearest the perpendicular, and that is indeed what we see.

When we look into the prism, we're seeing its true form, which is identical to a diffraction grating, going from blue to red. So, this matches the model, but we still must explain why, when the light is projected out onto a surface, the colours are reversed.

The "pinhole effect"

I suggest the reversal of colours is due to refraction creating a virtual "slit" or "pinhole" through which the original image is viewed.

Refraction causes the light exiting the surface to be "fanned out", as depicted in the diagram below. This creates an effect somewhat like a pinhole camera which inverts the original image, although it's not a circular hole, but a slit, so it only reverses in one dimension.

Also, it only reverses the colours, not the original image. I'll try to explain the mechanism.

When projecting the light onto a horizontal surface, we stand the prism up so it's vertical. In this case, refraction causes the original image to appear as if it's behind a horizontal slit or "doorway" when we look through it.

The overall effect is a combination of diffraction producing the colours, and refraction fanning them out, thus inverting them. (The eye in the diagram below would be to the right, looking down and into the glass.)

Observation

If we look at a single dot of white light coming through the prism, it is "smeared" in one dimension, into a line, with it tending to blue at one end and red at the other. So, a dot of white becomes a line of colours.

Fundamentally, the line of colours is wider than the dot. This "widening" of the image causes the edges to be obscured at distance.

If we move away from the prism, it starts to appear smaller than the image behind it, so the edges of the image are "clipped". If the eye is far enough back, we can only see the white light in the middle and the coloured ends of the spectrum are hidden.

If we move the eye side to side (or up/down), we can see that the spectrum appears to be behind an opening. We must move the eye to see one end of the spectrum or the other.

This is the effect that causes the colours to be reversed when they are projected out onto a surface.

A doorway

This reversal of colour (but not the underlying image) produced by the prism is like the following situation.

Imagine you're standing directly in front of a doorway (that has no door in it to get in the way). It's about 12 feet (4m) away. Behind the wall to the left of the doorway is a lamp with a red bulb, maybe 2ft back. You can't see the lamp from where you're standing, but you can see the red light it casts coming out of the doorway in front of you and going past to your right.

Now imagine a blue light behind the wall on the other side of the doorway, casting its light out to your left.

The two coloured beams will cross over exactly as they go through the doorway, making a X shape.

Because of the "fanning out" caused by refraction, the prism looks like the doorway. It puts the red and blue lights to the left and right behind it, out of sight, so they project out to the opposite side.

This is effectively what's going on with the prism. The red lamp is actually to the left, but the "wall" is preventing you from being able to see it directly, so its light is projected out to your right. The two beams appear to cross exactly at the surface / doorway.

I hope that makes sense.

The overall effect that the prism produces is thus a combination of diffraction and refraction happening together. This model seems to account for all its observed behaviours.

I apologise that this topic could use more explanation, diagrams and so on, but these effects are very easy to replicate if you have a prism. I hope you'll forgive me for leaving it here.

The End

Dear reader, thank you for making it all the way through the book. I'm grateful for your time and attention.

I hope you found the investigation as enlightening as I did.

Hopefully, the UP will lead us in the direction it indicates.

Table of Contents

Start	1
Introduction	4
A Universal Theory of Everything	4
Overview	14
Some Definitions	24
Duality	36
The Origin of Information	36
Two Types of Duality	36
One / Many	36
Measurement Creates Quantities	36
The Spirit / Matter Duality	53
The Unitary and Multiple Sets	53
Inside / Outside	53
Time	68
Classifying Dualities	68
Unity	81
One	81
Reflection	81
The Purpose of The Universe	81
The Mirror	81
The Footballer Analogy	81

Defining Logic	81
The Belly of the Whole	90
The Shadow Analogy	90
The Paradox of Yin	90

Four Elements — 100

Two Times Two	100
A Process of Creation	100
Alchemical Elements	100
The Human Body	100
Other Fours	111
The Computer Analogy	111
Arrangements of Four	111
Four Senses	124
Self-Categorisation	124

Seven Principles — 132

"Voice", "Heart" and "Sex"	132
The Three Operators	132
Voice: "Breathing Fire"	132
Heart: Cycles	144
Sex: Creation	144
The Archetype of Genetics?	144
Parts of Mind	144
Conclusion	144

Language — 152

The Parts of Speech	152
A Map of Language	152

The Heart — 166

Mechanisms and Machines	166
The Heart Mechanism	166
The Blood	166
The Third Person View	166

Applications — 182

Light, Energy, and Matter	182
The Mindlike Machine	208
Morality	223
Proof of God	235
Free Will	248
Alchemy	274
Final Thoughts	285

A New Theory of Colour — 295

Introduction	295
Current Theory	295
Observation of Prisms	295
Four-Primary Colour Theory	312
The Spectrum	312
Systole / Diastole	312
Conclusion	312
A New Theory of The Prism	328

www.ingramcontent.com/pod-product-compliance
Lightning Source LLC
Chambersburg PA
CBHW051146290426
44108CB00019B/2626